5G Mobile Networks:
A Systems Approach

Synthesis Lectures on Network Systems

Editor
Larry Peterson, *Open Networking Foundation and Princeton University*

Synthesis Lectures on Network Systems is edited by Larry Peterson, CTO of the Open Networking Foundation and Robert E. Kahn Professor Emeritus at Princeton University. The series publishes 75- to 200-page books covering a systems approach to networking and the distributed computing systems built around networks. The major emphasis is placed on the design and implementation of networks, distributed systems, and scalable cloud services in order to address a combination of traditional networking topics (e.g., congestion control, access control, addressing/routing, virtualization, real-time streaming, and mobility) and universal systems concerns (e.g., scalability, reliability, availability, security, feature velocity, and manageability). Although not a strict requirement, the series prioritizes empirical results, real-world use cases, and experience building systems. The scope will largely follow the purview of premier networking and systems conferences, such as SIGCOMM, NSDI, SOSP, and OSDI.

5G Mobile Networks: A Systems Approach

Larry Peterson and Oğuz Sunay

www.morganclaypool.com

ISBN: 9781681738888   paperback
ISBN: 9781681738895   ebook
ISBN: 9781681738901   hardcover

DOI 10.2200/S01021ED1V01Y202006NSY001

A Publication in the Morgan & Claypool Publishers series
SYNTHESIS LECTURES ON NETWORK SYSTEMS

Lecture #1
Series Editor: Larry Peterson, *Open Networking Foundation and Princeton University*
ISSN pending.

5G Mobile Networks: A Systems Approach

Larry Peterson

Open Networking Foundation and Princeton University

Oğuz Sunay

Open Networking Foundation

SYNTHESIS LECTURES ON NETWORK SYSTEMS #1

MORGAN & CLAYPOOL PUBLISHERS

ABSTRACT

This book describes the 5G mobile network from a systems perspective, focusing on the fundamental design principles that are easily obscured by an overwhelming number of acronyms and standards definitions that dominate this space. The book is written for system generalists with the goal of helping bring up to speed a community that understands a broad range of systems issues (but knows little or nothing about the cellular network) so it can play a role in the network's evolution. This is a community that understands both feature velocity and best practices in building robust scalable systems, and so it has an important role to play in bringing to fruition all of 5G's potential.

In addition to giving a step-by-step tour of the design rationale behind 5G, the book aggressively disaggregates the 5G mobile network. Building a disaggregated, virtualized, and software-defined 5G access network is the direction the industry is already headed (for good technical and business reasons), but breaking the 5G network down into its elemental components is also the best way to explain how 5G works. It also helps to illustrate how 5G might evolve in the future to provide even more value.

An open source implementation of 5G serves as the technical underpinning for the book. The authors, in collaboration with industrial and academic partners, are working towards a cloud-based implementation that takes advantage of both Software-Defined Networking (SDN) and cloud-native (microservice-based) architectures, culminating in a managed 5G-enabled EdgeCloud-as-a-Service built on the components and mechanisms described throughout the book.

KEYWORDS

5G, mobile networks, edge cloud, SDN, open source networking software

Contents

Preface

The transition to 5G is happening, and unless you've been actively trying to ignore it, you've un-doubtedly heard the hype. But if you are like 99% of the CS-trained, systems-oriented, cloud-savvy people in the world, the cellular network is largely a mystery. You know it's an important technology used in the last mile to connect people to the Internet, but you've otherwise abstracted it out of your scope-of-concerns.

The important thing to understand about 5G is that it implies much more than a generational upgrade in bandwidth. It involves transformative changes that blur the line between the access network and the cloud. And it will encompass enough value that it has the potential to turn the "Access-as-frontend-to-Internet" perspective on its head. We are just as likely to be talking about "Internet-as-backend-to-Access" ten years from now.

This book is written for someone that has a working understanding of the Internet and cloud, but has had limited success penetrating the myriad of acronyms that dominate cellular networking. In fairness, the Internet has its share of acronyms, but it also comes with a sufficient set of abstractions to help manage the complexity. It's hard to say the same for the cellular network, where pulling on one thread seemingly unravels the entire space. It has also been the case that the cellular network had been largely hidden inside proprietary devices, which has made it impossible to figure it out for yourself.

This book is the result of a mobile networking expert teaching a systems person about 5G as we've collaborated on an open source 5G implementation. The material has been used to train other software developers, and we are hopeful it will be useful to anyone that wants a deeper understanding of 5G and the opportunity for innovation it provides. Readers that want hands-on experience can also access the open source software introduced in the book.

This book will likely be a work-in-progress for the foreseeable future. It's not intended to be encyclopedic—favoring perspective and end-to-end completeness over every last bit of detail—but we do plan to flesh out the content over time. Your suggestions (and contributions) to this end are welcome.

Larry Peterson and Oğuz Sunay
Open Networking Foundation
June 2020

CHAPTER 1

Introduction

Mobile networks, which have a 40-year history that parallels the Internet's, have undergone significant change. The first two generations supported voice and then text, with 3G defining the transition to broadband access, supporting data rates measured in hundreds of kilobits-per-second. Today, the industry is at 4G (supporting data rates typically measured in the few megabits-per-second) and starting the transition to 5G, with the promise of a tenfold increase in data rates.

But 5G is about much more than increased bandwidth. In the same way 3G defined the transition from voice to broadband, 5G's promise is primarily about the transition from a single access service (broadband connectivity) to a richer collection of edge services and devices, including support for immersive user interfaces (e.g., AR/VR), mission-critical applications (e.g., public safety, autonomous vehicles), and the Internet-of-Things (IoT). Because these use cases will include everything from home appliances to industrial robots to self-driving cars, 5G won't just support humans accessing the Internet from their smartphones, but also swarms of autonomous devices working together on their behalf. All of this requires a fundamentally different architecture.

The requirements for this architecture are ambitious, and can be summarized as having three main objectives.

- To support *Massive Internet-of-Things*, potentially including devices with ultra-low energy (10+ years of battery life), ultra-low complexity (10s of bits-per-second), and ultra-high density (1 million nodes per square kilometer).

- To support *Mission-Critical Control*, potentially including ultra-high availability (greater than 10^{-5} per ms), ultra-low latency (as low as 1 ms), and extreme mobility (up to 100 km/h).

- To support *Enhanced Mobile Broadband*, potentially including extreme capacity (10 Tbps per square kilometer) and extreme data rates (multi-Gbps peak, 100+ Mbps sustained).

These targets will certainly not be met overnight, but that's in keeping with each generation of the mobile network being a decade-long endeavor.

Further Reading

For an example of the grand vision for 5G from one of the industry leaders, see Making 5G NR a Reality. Qualcomm Whitepaper, December 2016.

The 5G mobile network, because it is on an evolutionary path and not a point solution, includes standardized specifications, a range of implementation choices, and a long list of aspirational goals. Because this leaves so much room for interpretation, our approach to describing 5G is grounded in two mutually supportive principles. The first is to apply a *systems lens*, which is to say, we explain the sequence of design decisions that lead to a solution rather than fall back on enumerating the overwhelming number of acronyms as a *fait accompli*. The second is to aggressively disaggregate the system. Building a disaggregated, virtualized, and software-defined 5G access network is the direction the industry is already headed (for good technical and business reasons), but breaking the 5G network down into its elemental components is also the best way to explain how 5G works. It also helps to illustrate how 5G might evolve in the future to provide even more value.

Evolutionary Path

That 5G is on an evolutionary path is the central theme of this book. We call attention to its importance here, and revisit the topic throughout the book.

We are writing this book for *system generalists*, with the goal of helping bring a community that understands a broad range of systems issues (but knows little or nothing about the cellular network) up to speed so they can play a role in its evolution. This is a community that understands both feature velocity and best practices in building robust scalable systems, and so has an important role to play in bringing all of 5G's potential to fruition.

What this all means is that there is no single, comprehensive definition of 5G, any more than there is for the Internet. It is a complex and evolving system, constrained by a set of standards that purposely give all the stakeholders many degrees of freedom. In the chapters that follow, it should be clear from the context whether we are talking about *standards* (what everyone must do to interoperate), *trends* (where the industry seems to be headed), or *implementation choices* (examples to make the discussion more concrete). By adopting a systems perspective throughout, our intent is to describe 5G in a way that helps the reader navigate this rich and rapidly evolving system.

1.1   STANDARDIZATION LANDSCAPE

As of 3G, the generational designation corresponds to a standard defined by the *3rd Generation Partnership Project (3GPP)*. Even though its name has "3G" in it, the 3GPP continues to define the standards for 4G and 5G, each of which corresponds to a sequence of releases of the standard. Release 15 is considered the demarcation point between 4G and 5G, with Release 17 scheduled for 2021. Complicating the terminology, 4G was on a multi-release evolutionary path referred to as *Long Term Evolution (LTE)*. 5G is on a similar evolutionary path, with several expected releases over its lifetime.

While 5G is an ambitious advance beyond 4G, it is also the case that understanding 4G is the first step to understanding 5G, as several aspects of the latter can be explained as bringing a new degree-of-freedom to the former. In the chapters that follow, we often introduce some architectural feature of 4G as a way of laying the foundation for the corresponding 5G component.

Like Wi-Fi, cellular networks transmit data at certain bandwidths in the radio spectrum. Unlike Wi-Fi, which permits anyone to use a channel at either 2.4 or 5 GHz (these are unlicensed bands), governments have auctioned off and licensed exclusive use of various frequency bands to service providers, who in turn sell mobile access service to their subscribers.

There is also a shared-license band at 3.5 GHz, called *Citizens Broadband Radio Service (CBRS)*, set aside in North America for cellular use. Similar spectrum is being set aside in other countries. The CBRS band allows three tiers of users to share the spectrum: first right of use goes to the original owners of this spectrum, naval radars and satellite ground stations; followed by priority users who receive this right over 10 MHz bands for 3 years via regional auctions; and finally the rest of the population, who can access and utilize a portion of this band as long as they first check with a central database of registered users. CBRS, along with standardization efforts to extend cellular networks to operate in the unlicensed bands, open the door for private cellular networks similar to Wi-Fi.

The specific frequency bands that are licensed for cellular networks vary around the world, and are complicated by the fact that network operators often simultaneously support both old/legacy technologies and new/next-generation technologies, each of which occupies a different frequency band. The high-level summary is that traditional cellular technologies range from 700–2400-MHz, with new mid-spectrum allocations now happening at 6 GHz, and millimeter-wave (mmWave) allocations opening above 24 GHz.

While the specific frequency band is not directly relevant to understanding 5G from an architectural perspective, it does impact the physical-layer components, which in turn has indirect ramifications on the overall 5G system. We identify and explain these ramifications in later chapters.

1.2   ACCESS NETWORKS

The cellular network is part of the access network that implements the Internet's so-called *last mile*. Other access technologies include *Passive Optical Networks (PON)*, colloquially known as Fiber-to-the-Home. These access networks are provided by both big and small network operators. Global network operators like AT&T run access networks at thousands of aggregation points-of-presence across a country like the U.S., along with a national backbone that interconnects those sites. Small regional and municipal network operators might run an access network with one or two points-of-presence, and then connect to the rest of the Internet through some large operator's backbone.

In either case, access networks are physically anchored at thousands of aggregation points-of-presence within close proximity to end users, each of which serves anywhere from 1,000–100,000 subscribers, depending on population density. In practice, the physical deployment of these "edge" locations vary from operator to operator, but one possible scenario is to anchor both the cellular and wireline access networks in Telco *Central Offices*.

Historically, the Central Office—officially known as the *PSTN (Public Switched Telephone Network) Central Office*—anchored wired access (both telephony and broadband), while the cellular network evolved independently by deploying a parallel set of *Mobile Telephone Switching Offices (MTSO)*. Each MTSO serves as a *mobile aggregation* point for the set of cell towers in a given geographic area. For our purposes, the important idea is that such aggregation points exist, and it is reasonable to think of them as defining the edge of the operator-managed access network. For simplicity, we sometimes use the term "Central Office" as a synonym for both types of edge sites.

1.3   EDGE CLOUD

Because of their wide distribution and close proximity to end users, Central Offices are also an ideal place to host the edge cloud. But this begs the question: what exactly is the edge cloud?

In a nutshell, the cloud began as a collection of warehouse-sized datacenters, each of which provided a cost-effective way to power, cool, and operate a scalable number of servers. Over time, this shared infrastructure lowered the barrier to deploying scalable Internet services, but today, there is increasing pressure to offer low-latency/high-bandwidth cloud applications that cannot be effectively implemented in centralized datacenters. Augmented Reality (AR), Virtual Reality (VR), Internet-of-Things (IoT), and Autonomous Vehicles are all examples of this kind of application. This has resulted in a trend to move some functionality out of the datacenter and towards the edge of the network, closer to end users.

Where this edge is *physically* located depends on who you ask. If you ask a network operator that already owns and operates thousands of Central Offices, then their Central Offices are an obvious answer. Others might claim the edge is located at the 14,000 Starbucks across the U.S., and still others might point to the tens-of-thousands of cell towers spread across the globe.

Our approach is to be location agnostic, but it is worth pointing out that the cloud's migration to the edge coincides with a second trend, which is that network operators are re-architecting the access network to use the same commodity hardware and best practices in building scalable software as the cloud providers. Such a design, which is sometimes referred to as *Central Office Re-architected as a Datacenter*, supports both the access network and edge services co-located on a shared cloud platform. This platform is then replicated across hundreds or thousands of sites (including, but not limited to, Central Offices). So while we shouldn't limit ourselves to the Central Office as the only answer to the question of where the edge cloud is located, it is becoming a viable option.

> ### Further Reading
>
> To learn about the technical origins of CORD, which was first applied to fiber-based access networks (PON), see Central Office Re-architected as a Datacenter, IEEE Communications, October 2016.
>
> To understand the business case for CORD (and CORD-inspired technologies), see the A.D. Little report Who Dares Wins! How Access Transformation Can Fast-Track Evolution of Operator Production Platforms, September 2019.

When we get into the details of how 5G can be implemented in practice, we use CORD as our exemplar. For now, the important thing to understand is that 5G is being implemented as software running on commodity hardware, rather than embedded in the special-purpose proprietary hardware used in past generations. This has a significant impact on how we think about 5G (and how we describe 5G), which will increasingly become yet another software-based component in the cloud, as opposed to an isolated and specialized technology attached to the periphery of the cloud.

Keep in mind that our use of CORD as an exemplar is not to imply that the edge cloud is limited to Central Offices. CORD is a good exemplar because it is designed to host both edge services and access technologies like 5G on a common platform, where the Telco Central Office is one possible location to deploy such a platform.

An important takeaway from this discussion is that to understand how 5G is being implemented, it is helpful to have a working understanding of how clouds are built. This includes the use of *commodity hardware* (both servers and white-box switches), horizontally scalable *microservices* (also referred to as *cloud native*), and *Software-Defined Networks (SDN)*. It is also helpful to have an appreciation for how cloud software is developed, tested, deployed, and operated, including practices like *DevOps* and *Continuous Integration/Continuous Deployment (CI/CD)*.

> ### Further Reading
>
> If you are unfamiliar with SDN, we recommend a companion book: Software-Defined Networks: A Systems Approach. March 2020.
>
> If you are unfamiliar with DevOps—or more generally, with the operational issues cloud providers face—we recommend Site Reliability Engineering: How Google Runs Production Systems.

One final note about terminology. Anyone that has been paying attention to the discussion surrounding 5G will have undoubtedly heard about *Network Function Virtualization (NFV)*, which involves moving functionality that was once embedded in hardware appliances into VMs running on commodity servers. In our experience, NFV is a stepping stone towards the fully disaggregated and cloud native solution we describe in this book, and so we do not dwell on it. You can think of the NFV initiative as mostly consistent with the approach taken in this book, but making some different engineering choices when we get down into the specifics of the implementation (e.g., NFV is generally VM-based rather than microservice-based).

Although equating NFV with a different implementation choice is perfectly valid, there is another interpretation of events that better captures the essence of the transformation currently underway. When Telcos began the NFV initiative, they imagined incorporating cloud technologies into their networks, creating a so-called *Telco Cloud*. What is actually happening instead, is that the Telco's access technology is being subsumed into the cloud, running as yet another cloud-hosted workload. It would be more accurate to refer to the resulting system now emerging as the *Cloud-based Telco*. One reading of this book is as a roadmap to such an outcome.

CHAPTER 2

Radio Transmission

For anyone familiar with wireless access technologies like Wi-Fi, the cellular network is most unique due to its approach to sharing the available radio spectrum among its many users, all the while allowing those users to remain connected while moving. This has resulted in a highly dynamic and adaptive approach, in which coding, modulation and scheduling play a central role.

As we will see in this chapter, cellular networks use a reservation-based strategy, whereas Wi-Fi is contention-based. This difference is rooted in each system's fundamental assumption about utilization: Wi-Fi assumes a lightly loaded network (and hence optimistically transmits when the wireless link is idle and backs off if contention is detected), while 4G and 5G cellular networks assume (and strive for) high utilization (and hence explicitly assign different users to different "shares" of the available radio spectrum).

We start by giving a short primer on radio transmission as a way of laying a foundation for understanding the rest of the 5G architecture. The following is not a substitute for a theoretical treatment of the topic, but is instead intended as a way of grounding the systems-oriented description of 5G that follows in the reality of wireless communication.

2.1 CODING AND MODULATION

The mobile channel over which digital data needs to be reliably transmitted brings a number of impairments, including noise, attenuation, distortion, fading, and interference. This challenge is addressed by a combination of coding and modulation, as depicted in Figure 2.1.

At its core, coding inserts extra bits into the data to help recover from all the environmental factors that interfere with signal propagation. This typically implies some form of *Forward Error Correction* (e.g., turbo codes, polar codes). Modulation then generates signals that represent the encoded data stream, and it does so in a way that matches the channel characteristics: it first uses a digital modulation signal format that maximizes the number of reliably transmitted bits every second based on the specifics of the observed channel impairments; it next matches the transmission bandwidth to channel bandwidth using pulse shaping; and finally, it uses RF modulation to transmit the signal as an electromagnetic wave over an assigned *carrier frequency*.

For a deeper appreciation of the challenges of reliably transmitting data by propagating radio signals through the air, consider the scenario depicted in Figure 2.2, where the signal bounces off various stationary and moving objects, following multiple paths from the transmitter to the receiver, who may also be moving.

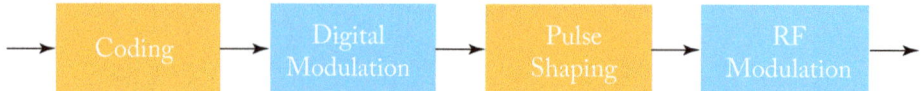

Figure 2.1: The role of coding and modulation in mobile communication.

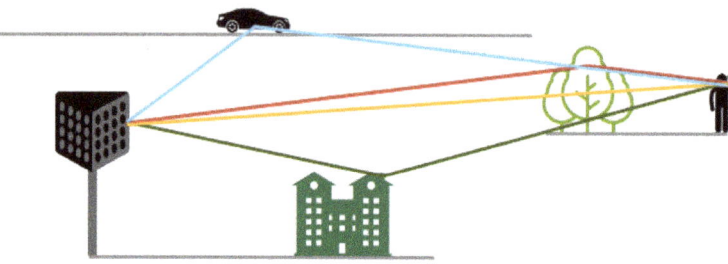

Figure 2.2: Signals propagate along multiple paths from transmitter to receiver.

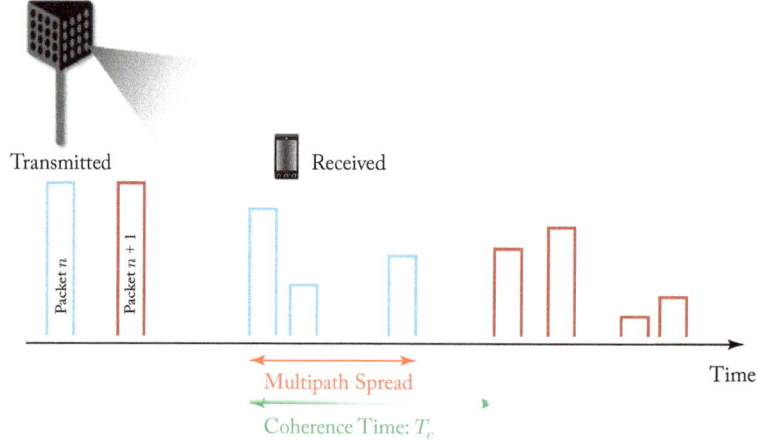

Figure 2.3: Received data spread over time due to multipath variation.

As a consequence of these multiple paths, the original signal arrives at the receiver spread over time, as illustrated in Figure 2.3. Empirical evidence shows that the Multipath Spread—the time between the first and last signals of one transmission arriving at the receiver—is 1–10 μs in urban environments and 10–30 μs in suburban environments. Theoretical bounds for the time duration for which the channel may be assumed to be time invariant, known as the *Coherence Time* and denoted T_c, is given by

$$T_c = c/v \times 1/f,$$

where c is the velocity of the signal, v is the velocity of the receiver (e.g., moving car or train), and f is the frequency of the carrier signal that is being modulated. This says the coherence time is inversely proportional to the frequency of the signal and the speed of movement, which makes intuitive sense: the higher the frequency (narrower the wave) the shorter the coherence time, and likewise, the faster the receiver is moving the longer the coherence time. Based on the target parameters to this model (selected according to the target physical environment), it is possible to calculate T_c, which in turn bounds the rate at which symbols can be transmitted without undue risk of interference.

To complicate matters further, Figures 2.2 and 2.3 imply the transmission originates from a single antenna, but cell towers are equipped with an array of antennas, each transmitting in a different (but overlapping) direction. This technology, called *Multiple-Input-Multiple-Output (MIMO)*, opens the door to purposely transmitting data from multiple antennas in an effort to reach the receiver, adding even more paths to the environment-imposed multipath propagation.

One of the most important consequences of these factors is that the transmitter must receive feedback from every receiver to judge how to best utilize the wireless medium on their behalf. 3GPP specifies a *Channel Quality Indicator (CQI)* for this purpose, where in practice the receiver sends a CQI status report to the base station periodically (e.g., every millisecond in LTE). These CQI messages report the observed signal-to-noise ratio, which impacts the receiver's ability to recover the data bits. The base station then uses this information to adapt how it allocates the available radio spectrum to the subscribers it is serving, as well as which coding and modulation scheme to employ. All of these decisions are made by the scheduler.

2.2   SCHEDULER

How the scheduler does its job is one of the most important properties of each generation of the cellular network, which in turn depends on the multiplexing mechanism. For example, 2G used *Time Division Multiple Access (TDMA)* and 3G used *Code Division Multiple Access (CDMA)*. How data is multiplexed is also a major differentiator for 4G and 5G, completing the transition from the cellular network being fundamentally circuit-switched to fundamentally packet-switched.

Both 4G and 5G are based on *Orthogonal Frequency-Division Multiplexing (OFDM)*, an approach that multiplexes data over multiple orthogonal subcarrier frequencies, each of which is modulated independently. The value and efficiency of OFDM is in how it selects subcarrier frequencies so as to avoid interference, that is, how it achieves orthogonality. That topic is beyond the scope of this book. We instead take a decidedly abstract perspective of multiplexing, focusing on "discrete scheduleable units of the radio spectrum" rather the the signalling and modulation underpinnings that yield those scheduleable units.

To start, we drill down on these schedulable units. We return to the broader issue of the *air interface* that makes efficient use of the spectrum in the concluding section.

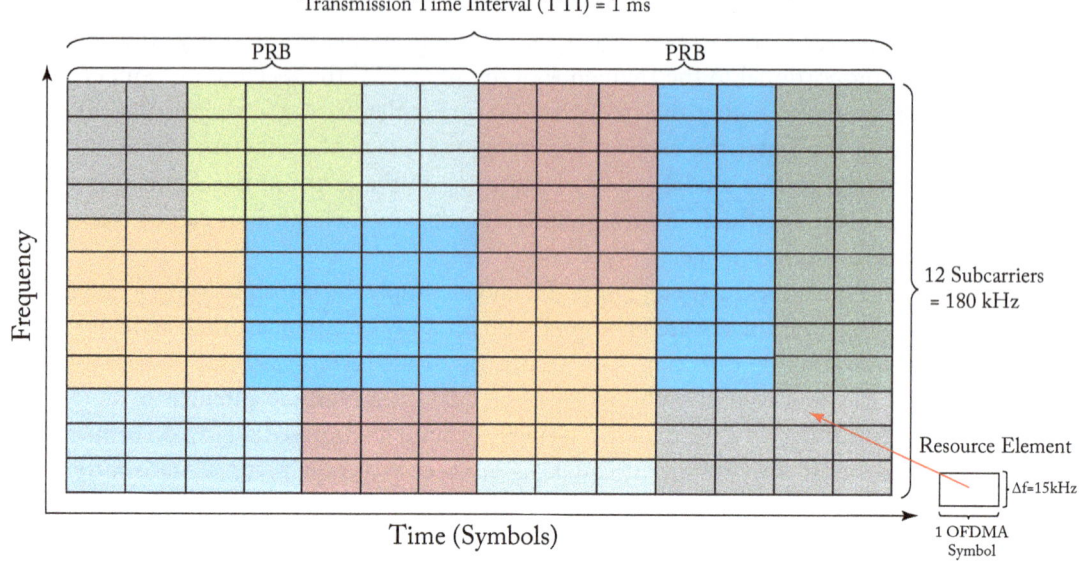

Figure 2.4: Spectrum abstractly represented by a 2-D grid of schedulable Resource Elements.

2.2.1   MULTIPLEXING IN 4G

The 4G approach to multiplexing downstream transmissions is called *Orthogonal Frequency-Division Multiple Access (OFDMA)*, a specific application of OFDM that multiplexes data over a set of 12 orthogonal subcarrier frequencies, each of which is modulated independently.[1] The "Multiple Access" in OFDMA implies that data can simultaneously be sent on behalf of multiple users, each on a different subcarrier frequency and for a different duration of time. The subbands are narrow (e.g., 15 kHz), but the coding of user data into OFDMA symbols is designed to minimize the risk of data loss due to interference between adjacent bands.

The use of OFDMA naturally leads to conceptualizing the radio spectrum as a 2-D resource, as shown in Figure 2.4. The minimal schedulable unit, called a *Resource Element (RE)*, corresponds to a 15-kHz-wide band around one subcarrier frequency and the time it takes to transmit one OFDMA symbol. The number of bits that can be encoded in each symbol depends on the modulation rate, so for example using *Quadrature Amplitude Modulation (QAM)*, 16-QAM yields 4 bits per symbol and 64-QAM yields 6 bits per symbol.

A scheduler allocates some number of REs to each user that has data to transmit during each 1 ms *Transmission Time Interval (TTI)*, where users are depicted by different colored blocks in Figure 2.4. The only constraint on the scheduler is that it must make its allocation decisions on blocks of $7 \times 12 = 84$ resource elements, called a *Physical Resource Block (PRB)*. Figure 2.4

---

[1]4G uses a different multiplexing strategy for upstream transmissions (from user devices to base stations), but we do not describe it because the approach is not applicable to 5G.

shows two back-to-back PRBs. Of course time continues to flow along one axis, and depending on the size of the available frequency band (e.g., it might be 100 MHz wide), there may be many more subcarrier slots (and hence PRBs) available along the other axis, so the scheduler is essentially preparing and transmitting a sequence of PRBs.

Note that OFDMA is not a coding/modulation algorithm, but instead provides a framework for selecting a specific coding and modulator for each subcarrier frequency. QAM is one common example modulator. It is the scheduler's responsibility to select the modulation to use for each PRB, based on the CQI feedback it has received. The scheduler also selects the coding on a per-PRB basis, for example, by how it sets the parameters to the turbo code algorithm.

The 1-ms TTI corresponds to the time frame in which the scheduler receives feedback from users about the quality of the signal they are experiencing. This is the CQI mentioned earlier, where once every millisecond, each user sends a set of metrics, which the scheduler uses to make its decision as to how to allocate PRBs during the subsequent TTI.

Another input to the scheduling decision is the *QoS Class Identifier (QCI)*, which indicates the quality-of-service each class of traffic is to receive. In 4G, the QCI value assigned to each class (there are nine such classes, in total) indicates whether the traffic has a *Guaranteed Bit Rate (GBR)* or not *(non-GBR)*, plus the class's relative priority within those two categories.

Finally, keep in mind that Figure 2.4 focuses on scheduling transmissions from a single antenna, but the MIMO technology described above means the scheduler also has to determine which antenna (or more generally, what subset of antennas) will most effectively reach each receiver. But again, in the abstract, the scheduler is charged with allocating a sequence of Resource Elements.

This all begs the question: how does the scheduler decide which set of users to service during a given time interval, how many resource elements to allocate to each such user, how to select the coding and modulation levels, and which antenna to transmit their data on? This is an optimization problem that, fortunately, we are not trying to solve here. Our goal is to describe an architecture that allows someone else to design and plug in an effective scheduler. Keeping the cellular architecture open to innovations like this is one of our goals, and as we will see in the next section, becomes even more important in 5G where the scheduler operates with even more degrees of freedom.

2.2.2   MULTIPLEXING IN 5G

The transition from 4G to 5G introduces additional flexibility in how the radio spectrum is scheduled, making it possible to adapt the cellular network to a more diverse set of devices and applications domains.

Fundamentally, 5G defines a family of waveforms—unlike LTE, which specified only one waveform—each optimized for a different band in the radio spectrum.[2] The bands with

---

[2]A waveform is the frequency, amplitude, and phase-shift independent property (shape) of a signal. A sine wave is an example waveform.

carrier frequencies below 1 GHz are designed to deliver mobile broadband and massive IoT services with a primary focus on range. Carrier frequencies between 1–6 GHz are designed to offer wider bandwidths, focusing on mobile broadband and mission-critical applications. Carrier frequencies above 24 GHz (mmWaves) are designed to provide super wide bandwidths over short, line-of-sight coverage.

These different waveforms affect the scheduling and subcarrier intervals (i.e., the "size" of the resource elements described in the previous section).

- For sub-1 GHz bands, 5G allows maximum 50-MHz bandwidths. In this case, there are two waveforms: one with subcarrier spacing of 15 kHz and another of 30 kHz. (We used 15 kHz in the example shown in Figure 2.4.) The corresponding scheduling intervals are 0.5 and 0.25 ms, respectively. (We used 0.5 ms in the example shown in Figure 2.4.)

- For 1–6 GHz bands, maximum bandwidths go up to 100 MHz. Correspondingly, there are three waveforms with subcarrier spacings of 15, 30, and 60 kHz, corresponding to scheduling intervals of 0.5, 0.25, and 0.125 ms, respectively.

- For millimeter bands, bandwidths may go up to 400 MHz. There are two waveforms, with subcarrier spacings of 60 kHz and 120 kHz. Both have scheduling intervals of 0.125 ms.

These various configurations of subcarrier spacing and scheduling intervals are sometimes called the *numerology* of the radio's air interface.

This range of numerology is important because it adds another degree of freedom to the scheduler. In addition to allocating radio resources to users, it has the ability to dynamically adjust the size of the resource by changing the wave form being used. With this additional freedom, fixed-sized REs are no longer the primary unit of resource allocation. We instead use more abstract terminology, and talk about allocating *Resource Blocks* to subscribers, where the 5G scheduler determines both the size and number of Resource Blocks allocated during each time interval.

Figure 2.5 depicts the role of the scheduler from this more abstract perspective, where just as with 4G, CQI feedback from the receivers and the QCI quality-of-service class selected by the subscriber are the two key pieces of input to the scheduler. Note that the set of QCI values changes between 4G and 5G, reflecting the increasing differentiation being supported. For 5G, each class includes the following attributes:

- Resource Type: Guaranteed Bit Rate (GBR), Delay-Critical GBR, Non-GBR

- Priority Level

- Packet Delay Budget

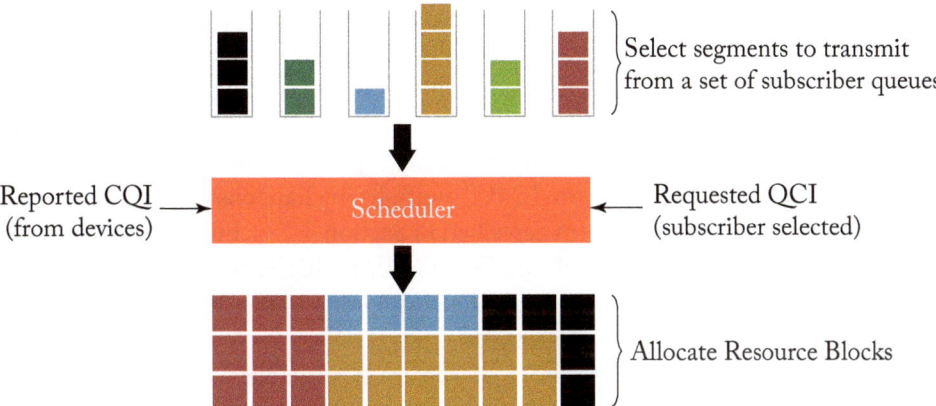
Select segments to transmit from a set of subscriber queues

Reported CQI (from devices) → Scheduler ← Requested QCI (subscriber selected)

Allocate Resource Blocks

Figure 2.5: Scheduler allocates Resource Blocks to user data streams based on CQI feedback from receivers and the QCI parameters associated with each class of service.

- Packet Error Rate

- Averaging Window

- Maximum Data Burst

Note that while the preceding discussion could be interpreted to imply a one-to-one relationship between subscribers and a QCI, it is more accurate to say that each QCI is associated with a class of traffic (often corresponding to some type of application), where a given subscriber might be sending and receiving traffic that belongs to multiple classes at any given time. We explore this idea in much more depth in a later chapter.

2.3   NEW RADIO (NR)

We conclude by noting that while the previous section describes 5G as introducing additional degrees of freedom into how data is scheduled for transmission, the end result is a qualitatively more powerful radio. This new 5G air interface specification, which is commonly referred to as *New Radio (NR)*, enables three new use cases that go well beyond simply delivering increased bandwidth:

- Extreme Mobile Broadband

- Ultra-Reliable Low-Latency Communications

- Massive Machine-Type Communications

All three correspond to the requirements introduced in Chapter 1, and can be attributed to four fundamental improvements in how 5G multiplexes data onto the radio spectrum.

The first is the one identified in the previous section: being able to change the waveform. This effectively introduces the ability to dynamically change the size and number of scheduleable resource units, which opens the door to making fine-grain scheduling decisions that are critical to predictable, low-latency communication.

The second is related to the "Multiple Access" aspect of how distinct traffic sources are multiplexed onto the available spectrum. In 4G, multiplexing happens in both the frequency and time domains for downstream traffic (as described in Section 2.2.1), but multiplexing happens in only the frequency domain for upstream traffic. 5G NR multiplexes both upstream and downstream traffic in both the time and frequency domains. Doing so provides finer-grain scheduling control needed by latency-sensitive applications.

The third is related to the plethora of spectrum available to 5G NR, with the new mmWave allocations opening above 24 GHz being especially important. This is not only because of the abundance of capacity—which makes it possible to set aside dedicated capacity for mission-critical applications that require low-latency communication—but also because the higher-frequency enables even finer-grain resource blocks (e.g., scheduling intervals as short as 0.125 ms). Again, this improves scheduling granularity to the benefit of applications that cannot tolerate unpredictable latency.

The fourth is related to delivering mobile connectivity to a massive number of IoT devices, ranging from devices that require mobility support and modest data rates (e.g. wearables, asset trackers) to devices that support intermittent transmission of a few bytes of data (e.g., sensors, meters). None of these devices are particularly latency-sensitive or bandwidth-greedy, but the latter are especially challenging because they require long battery lifetimes, and hence, reduced hardware complexity that draws less power.

Support for IoT device connectivity revolves around allocating some of the available radio spectrum to a light-weight (simplified) air interface. This approach started with Release 13 of LTE via two complementary technologies: mMTC and NB-IoT (NarrowBand-IoT). Both technologies build on a significantly simplified version of LTE—i.e., limiting the numerology and flexibility needed achieve high spectrum utilization—so as to allow for simpler IoT hardware design. mMTC delivers up to 1 Mbps over a 1.4 MHz of bandwidth and NB-IoT delivers a few tens of kbps over 200 kHz of bandwidth; hence the term *NarrowBand*. Both technologies have been designed to support over 1 million devices per square kilometer. With Release 16, both technologies can be operated in-band with 5G, but still based on LTE numerology. Starting with Release 17, a simpler version of 5G NR, called *NR-Light*, will be introduced as the evolution of mMTC. NR-Light is expected to scale the device density even further.

As a consequence of all four improvements, 5G NR is designed to support partitioning the available bandwidth, with different partitions dynamically allocated to different classes of traffic (e.g., high-bandwidth, low-latency, and low-complexity). This is the essence of *slicing*, an idea we will revisit throughout this book. Moreover, once traffic with different requirements can be served by different slices, 5G NR's approach to multiplexing is general enough to support varied scheduling decisions for those slices, each tailored for the target traffic.

CHAPTER 3

Basic Architecture

This chapter identifies the main architectural components of cellular access networks. It focuses on the components that are common to both 4G and 5G and, as such, establishes a foundation for understanding the advanced features of 5G presented in later chapters.

This overview is partly an exercise in introducing 3GPP terminology. For someone that is familiar with the Internet, this terminology can seem arbitrary (e.g., "eNB" is a "base station"), but it is important to keep in mind that this terminology came out of the 3GPP standardization process, which has historically been concerned about telephony and almost completely disconnected from the IETF and other Internet-related efforts. To further confuse matters, the 3GPP terminology often changes with each generation (e.g., a base station is called eNB in 4G and gNB in 5G). We address situations like this by using generic terminology (e.g., base station), and referencing the 3GPP-specific counterpart only when the distinction is helpful.

Further Reading

This example is only the tip of the terminology iceberg. For a slightly broader perspective on the complexity of terminology in 5G, see Marcin Dryjanski's blog post: LTE and 5G Differences: System Complexity. July 2018.

3.1 MAIN COMPONENTS

The cellular network provides wireless connectivity to devices that are on the move. These devices, which are known as *User Equipment (UE)*, have until recently corresponded to smartphones and tablets, but will increasingly include cars, drones, industrial and agricultural machines, robots, home appliances, medical devices, and so on.

As shown in Figure 3.1, the cellular network consists of two main subsystems: the *Radio Access Network (RAN)* and the *Mobile Core*. The RAN manages the radio spectrum, making sure it is both used efficiently and meets the quality-of-service requirements of every user. The main component in the RAN is the cryptically named *eNodeB* (or *eNB*), which is short for the equally cryptic *evolved Node B*. The Mobile Core is a bundle of functionality (as opposed to a device) that serves several purposes.

- Provides Internet (IP) connectivity for both data and voice services.

- Ensures this connectivity fulfills the promised QoS requirements.

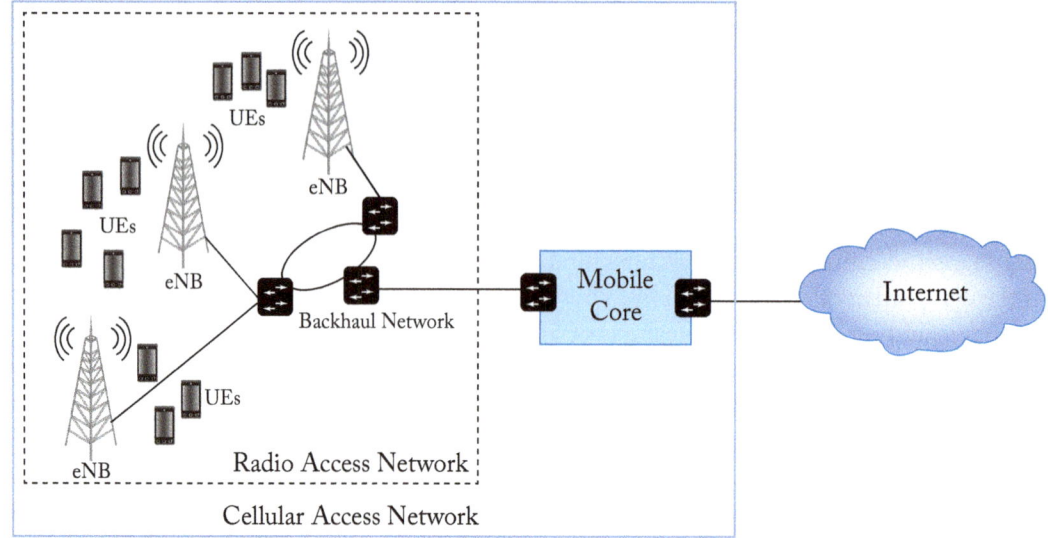

Figure 3.1: Cellular networks consists of a Radio Access Network (RAN) and a Mobile Core.

- Tracks user mobility to ensure uninterrupted service.

- Tracks subscriber usage for billing and charging.

Note that Mobile Core is another example of a generic term. In 4G this is called the *Evolved Packet Core (EPC)* and in 5G it is called the *Next Generation Core (NG-Core)*.

Even though the word "Core" is in its name, from an Internet perspective, the Mobile Core is still part of the access network, effectively providing a bridge between the RAN in some geographic area and the greater IP-based Internet. 3GPP provides significant flexibility in how the Mobile Core is geographically deployed, for our purposes, assuming each instantiation of the Mobile Core serves a metropolitan area is a good working model. The corresponding RAN would then span several dozens (or even hundreds) of cell towers.

Taking a closer look at Figure 3.1, we see that a *Backhaul Network* interconnects the eNBs that implement the RAN with the Mobile Core. This network is typically wired, may or may not have the ring topology shown in the Figure, and is often constructed from commodity components found elsewhere in the Internet. For example, the PON that implements Fiber-to-the-Home is a prime candidate for implementing the RAN backhaul. The backhaul network is obviously a necessary part of the RAN, but it is an implementation choice and not prescribed by the 3GPP standard.

Although 3GPP specifies all the elements that implement the RAN and Mobile Core in an open standard—including sub-layers we have not yet introduced—network operators have historically bought proprietary implementations of each subsystem from a single vendor. This

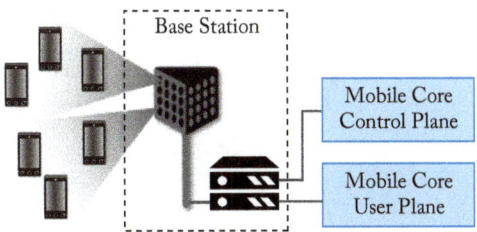

Figure 3.2: Mobile Core divided into a Control Plan and a User Plane, an architectural feature known as CUPS: Control and User Plane Separation.

lack of an open source implementation contributes to the perceived "opaqueness" of the cellular network in general, and the RAN in particular. And while it is true that an eNodeB implementation does contain sophisticated algorithms for scheduling transmission on the radio spectrum—algorithms that are considered valuable Intellectual Property of the equipment vendors—there is significant opportunity to open and disaggregate both the RAN and the Mobile Core. The following two sections describe each, in turn.

Before getting to those details, Figure 3.2 redraws components from Figure 3.1 to highlight two important distinctions. The first is that the eNB (which we will refer to as the Base Station from here on) has an analog component (depicted by an antenna) and a digital component (depicted by a processor pair). The second is that the Mobile Core is partitioned into a *Control Plane* and *User Plane*, which is similar to the control/data plane split that someone familiar with the Internet would recognize. (3GPP also recently introduced a corresponding acronym—*CUPS, Control and User Plane Separation*—to denote this idea.) The importance of these two distinctions will become clear in the following discussion.

3.2 RADIO ACCESS NETWORK

We now describe the RAN by sketching the role each base station plays. Keep in mind this is kind of like describing the Internet by explaining how a router works—a not unreasonable place to start, but it doesn't fully do justice to the end-to-end story.

First, each base station establishes the wireless channel for a subscriber's UE upon power-up or upon handover when the UE is active. This channel is released when the UE remains idle for a predetermined period of time. Using 3GPP terminology, this wireless channel is said to provide a bearer service.[1]

Second, each base station establishes "3GPP Control Plane" connectivity between the UE and the corresponding Mobile Core Control Plane component, and forwards signaling traffic

---

[1] The term "bearer" has historically been used in telecommunications (including early wireline technologies like ISDN) to denote a data channel, as opposed to a channel that carries signaling information.

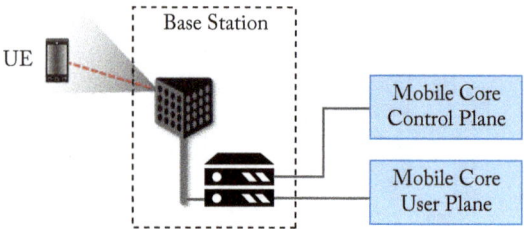

Figure 3.3: Base Station detects (and connects to) active UEs.

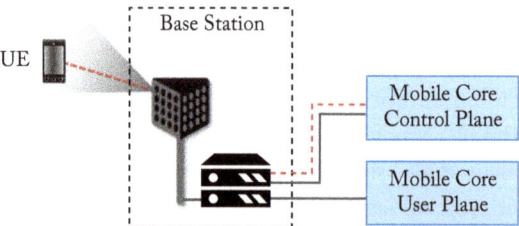

Figure 3.4: Base Station establishes control plane connectivity between each UE and the Mobile Core.

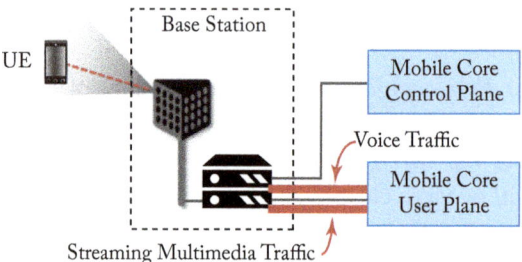

Figure 3.5: Base station establishes one or more tunnels between each UE and the Mobile Core's User Plane.

between the two. This signaling traffic enables UE authentication, registration, and mobility tracking.

Third, for each active UE, the base station establishes one or more tunnels between the corresponding Mobile Core User Plane component.

Fourth, the base station forwards both control and user plane packets between the Mobile Core and the UE. These packets are tunnelled over SCTP/IP and GTP/UDP/IP, respectively. SCTP (Stream Control Transport Protocol) is 3GPP-defined alternative to TCP, tailored to carry signaling (control) information for telephony services. GTP (a nested acronym corre-

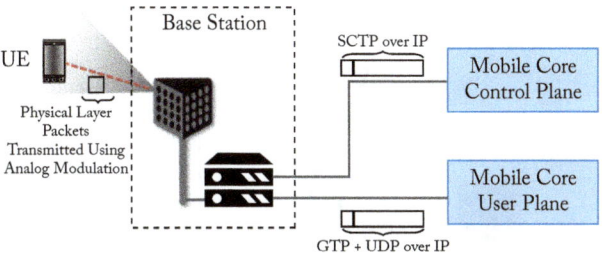

Figure 3.6: Base Station to Mobile Core (and Base Station to Base Station) control plane tunneled over SCTP/IP and user plane tunneled over GTP/UDP/IP.

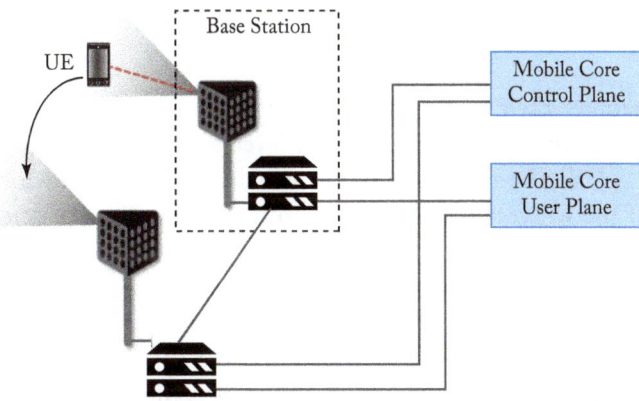

Figure 3.7: Base Stations cooperate to implement UE hand over.

sponding to (General Packet Radio Service) Tunneling Protocol) is a 3GPP-specific tunneling protocol designed to run over UDP.

As an aside, it is noteworthy that connectivity between the RAN and the Mobile Core is IP-based. This was introduced as one of the main changes between 3G and 4G. Prior to 4G, the internals of the cellular network were circuit-based, which is not surprising given its origins as a voice network.

Fifth, the base station coordinates UE handovers between neighboring base stations, using direct station-to-station links. Exactly like the station-to-core connectivity, shown in Figure 3.6, these links are used to transfer both control plane (SCTP over IP) and user plane (GTP over UDP/IP) packets.

Sixth, the base station coordinates wireless multi-point transmission to a UE from multiple base stations, which may or may not be part of a UE handover from one base station to another.

For our purposes, the main takeaway is that the base station can be viewed as a specialized forwarder. In the Internet-to-UE direction, it fragments outgoing IP packets into physical layer

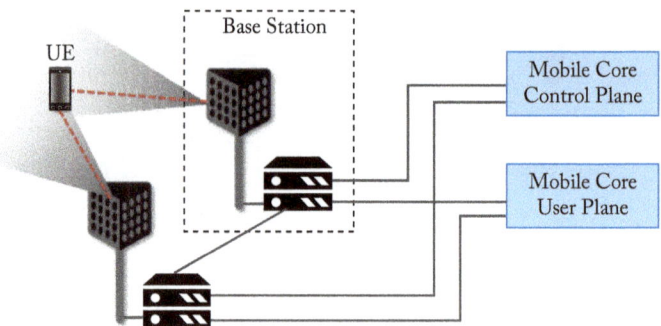

Figure 3.8: Base Stations cooperate to implement multipath transmission (link aggregation) to UEs.

segments and schedules them for transmission over the available radio spectrum, and in the UE-to-Internet direction it assembles physical layer segments into IP packets and forwards them (over a GTP/UDP/IP tunnel) to the upstream user plane of the Mobile Core. Also, based on observations of the wireless channel quality and per-subscriber policies, it decides whether to (a) forward outgoing packets directly to the UE, (b) indirectly forward packets to the UE via a neighboring base station, or (c) utilize multiple paths to reach the UE. The third case has the option of either spreading the physical payloads across multiple base stations or across multiple carrier frequencies of a single base station (including Wi-Fi).

Note that as outlined in Chapter 2, scheduling is complex and multi-faceted, even when viewed as a localized decision at a single base station. What we now see is that there is also a global element, whereby it's possible to forward traffic to a different base station (or to multiple base stations) in an effort to make efficient use of the radio spectrum over a larger geographic area.

In other words, the RAN as a whole (i.e., not just a single base station) not only supports handovers (an obvious requirement for mobility), but also *link aggregation* and *load balancing*, mechanisms that are familiar to anyone that understands the Internet. We will revisit how such RAN-wide (global) decisions can be made using SDN techniques in a later chapter.

3.3   MOBILE CORE

The main function of the Mobile Core is to provide external packet data network (e.g., Internet) connectivity to mobile subscribers, while ensuring that they are authenticated and their observed service qualities satisfy their subscription SLAs. An important aspect of the Mobile Core is that it needs to manage all subscribers' mobility by keeping track of their last whereabouts at the granularity of the serving base station.

While the aggregate functionality remains largely the same as we migrate from 4G to 5G, how that functionality is virtualized and factored into individual components changes, with the

5G Mobile Core heavily influenced by the cloud's march toward a microservice-based (cloud native) architecture. This shift to cloud native is deeper than it might first appear, in part because it opens the door to customization and specialization. Instead of supporting just voice and broadband connectivity, the 5G Mobile Core can evolve to also support, for example, massive IoT, which has a fundamentally different latency requirement and usage pattern (i.e., many more devices connecting intermittently). This stresses—if not breaks—a one-size-fits-all approach to session management.

3.3.1   4G MOBILE CORE

The 4G Mobile Core, which 3GPP officially refers to as the *Evolved Packet Core (EPC)*, consists of five main components, the first three of which run in the Control Plane (CP) and the second two of which run in the User Plane (UP).

- MME (Mobility Management Entity): Tracks and manages the movement of UEs throughout the RAN. This includes recording when the UE is not active.

- HSS (Home Subscriber Server): A database that contains all subscriber-related information.

- PCRF (Policy and Charging Rules Function): Tracks and manages policy rules and records billing data on subscriber traffic.

- SGW (Serving Gateway): Forwards IP packets to and from the RAN. Anchors the Mobile Core end of the bearer service to a (potentially mobile) UE, and so is involved in handovers from one base station to another.

- PGW (Packet Gateway): Essentially an IP router, connecting the Mobile Core to the external Internet. Supports additional access-related functions, including policy enforcement, traffic shaping, and charging.

Although specified as distinct components, in practice the SGW (RAN-facing) and PGW (Internet-facing) are often combined in a single device, commonly referred to as an S/PGW. The end result is illustrated in Figure 3.9.

Note that 3GPP is flexible in how the Mobile Core components are deployed to serve a geographic area. For example, a single MME/PGW pair might serve a metropolitan area, with SGWs deployed across ~10 edge sites spread throughout the city, each of which serves ~100 base stations. But alternative deployment configurations are allowed by the spec.

3.3.2   5G MOBILE CORE

The 5G Mobile Core, which 3GPP calls the *NG-Core*, adopts a microservice-like architecture, where we say "microservice-like" because while the 3GPP specification spells out this level of disaggregation, it is really just prescribing a set of functional blocks and not an implementation.

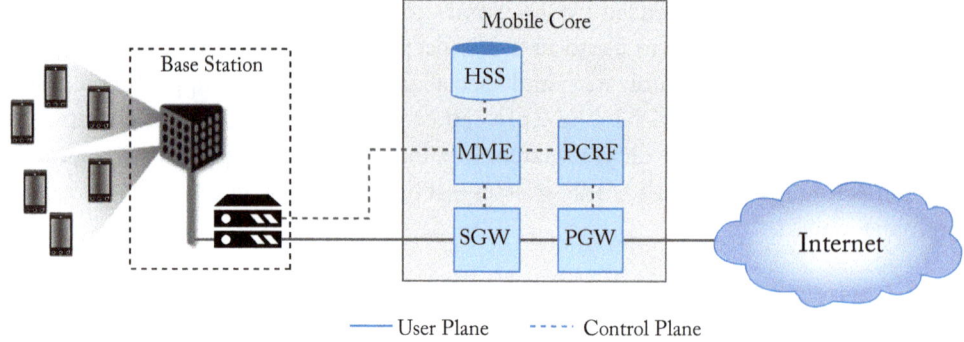

Figure 3.9: 4G Mobile Core (Evolved Packet Core).

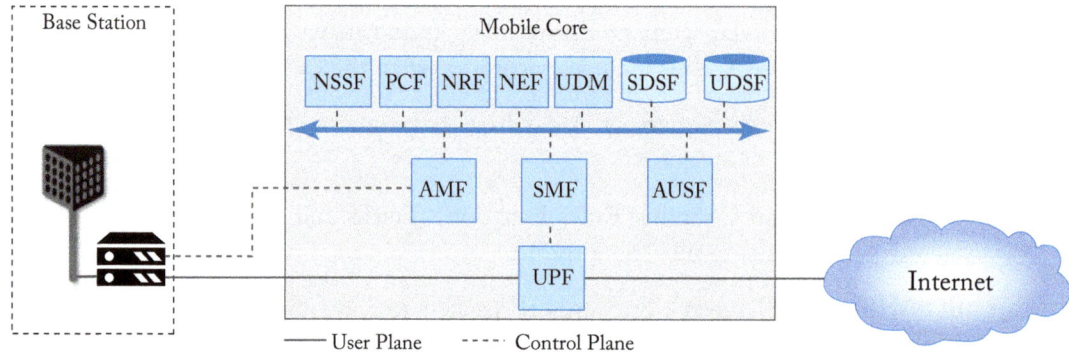

Figure 3.10: 5G Mobile Core (NG-Core).

Keeping in mind a set of functional blocks is very different from the collection of engineering decisions that go into designing a microservice-based system, viewing the collection of components shown in Figure 3.10 as a set of microservices is a good working model.

The following organizes the set of functional blocks into three groups. The first group runs in the Control Plane (CP) and has a counterpart in the EPC.

- AMF (Core Access and Mobility Management Function): Manages the mobility-related aspects of the EPC's MME. Responsible for connection and reachability management, mobility management, access authentication and authorization, and location services.

- SMF (Session Management Function): Manages each UE session, including IP address allocation, selection of associated UP function, control aspects of QoS, and control aspects of UP routing. Roughly corresponds to part of the EPC's MME and the control-related aspects of the EPC's PGW.

- PCF (Policy Control Function): Manages the policy rules that other CP functions then enforce. Roughly corresponds to the EPC's PCRF.

- UDM (Unified Data Management): Manages user identity, including the generation of authentication credentials. Includes part of the functionality in the EPC's HSS.

- AUSF (Authentication Server Function): Essentially an authentication server. Includes part of the functionality in the EPC's HSS.

The second group also runs in the Control Plane (CP) but does not have a direct counterpart in the EPC:

- SDSF (Structured Data Storage Network Function): A "helper" service used to store structured data. Could be implemented by an "SQL Database" in a microservices-based system.

- UDSF (Unstructured Data Storage Network Function): A "helper" service used to store unstructured data. Could be implemented by a "Key/Value Store" in a microservices-based system.

- NEF (Network Exposure Function): A means to expose select capabilities to third-party services, including translation between internal and external representations for data. Could be implemented by an "API Server" in a microservices-based system.

- NRF (NF Repository Function): A means to discover available services. Could be implemented by a "Discovery Service" in a microservices-based system.

- NSSF (Network Slicing Selector Function): A means to select a Network Slice to serve a given UE. Network slices are essentially a way to differentiate service given to different users. It is a key feature of 5G that we discuss in depth later in a later chapter.

The third group includes the one component that runs in the User Plane (UP):

- UPF (User Plane Function): Forwards traffic between RAN and the Internet, corresponding to the S/PGW combination in EPC. In addition to packet forwarding, it is responsible for policy enforcement, lawful intercept, traffic usage reporting, and QoS policing.

Of these, the first and third groups are best viewed as a straightforward refactoring of 4G's EPC, while the second group—despite the gratuitous introduction of new terminology—is 3GPP's way of pointing to a cloud native solution as the desired end-state for the Mobile Core. Of particular note, introducing distinct storage services means that all the other services can be stateless, and hence, more readily scalable. Also note that Figure 3.10 adopts an idea that's common in microservice-based systems, namely, to show a *message bus* interconnecting

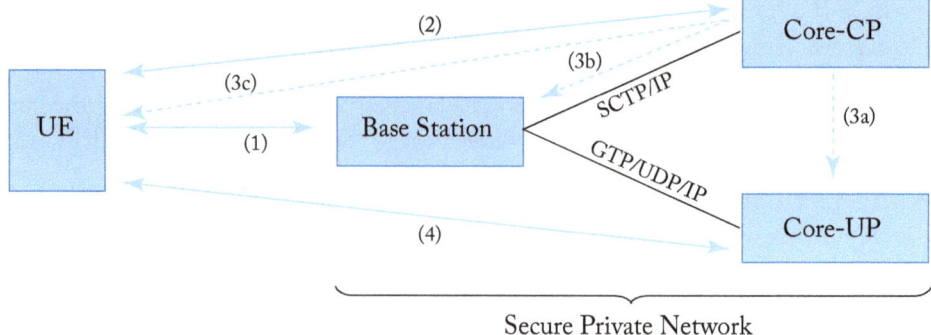

Figure 3.11: Sequence of steps to establish secure Control and User Plane channels.

all the components rather than including a full set of pairwise connections. This also suggests a well-understood implementation strategy.

Stepping back from these details, and with the caveat that we are presuming an implementation, the main takeaway is that we can conceptualize the Mobile Core as a *Service Mesh*. We adopt this terminology for "an interconnected set of microservices" since it is widely used in cloud native systems. Other terms you will sometimes hear are *Service Graph* and *Service Chain*, the latter being more prevalent in NFV-oriented documents. 3GPP is silent on the specific terminology since it is considered an implementation choice rather than part of the specification. We describe our implementation choices in later chapters.

3.4   SECURITY

We now take a closer look at the security architecture of the cellular network, which also serves to fill in some details about how each individual UE connects to the network. The architecture is grounded in two trust assumptions.

First, each Base Station trusts that it is connected to the Mobile Core by a secure private network, over which it establishes the tunnels introduced in Figure 3.6: a GTP/UDP/IP tunnel to the Core's User Plane (Core-UP) and a SCTP/IP tunnel to the Core's Control Plane (Core-CP). Second, each UE has an operator-provided SIM card, which uniquely identifies the subscriber (i.e., phone number) and establishes the radio parameters (e.g., frequency band) need to communicate with that operator's Base Stations. The SIM card also includes a secret key that the UE uses to authenticate itself.

With this starting point, Figure 3.11 shows the per-UE connection sequence. When a UE first becomes active, it communicates with a nearby Base Station over a temporary (unauthenticated) radio link (Step 1). The Base Station forwards the request to the Core-CP over the existing tunnel, and the Core-CP (specifically, the MME in 4G and the AMF in 5G) initiates an authentication protocol with the UE (Step 2). 3GPP identifies a set of options, including

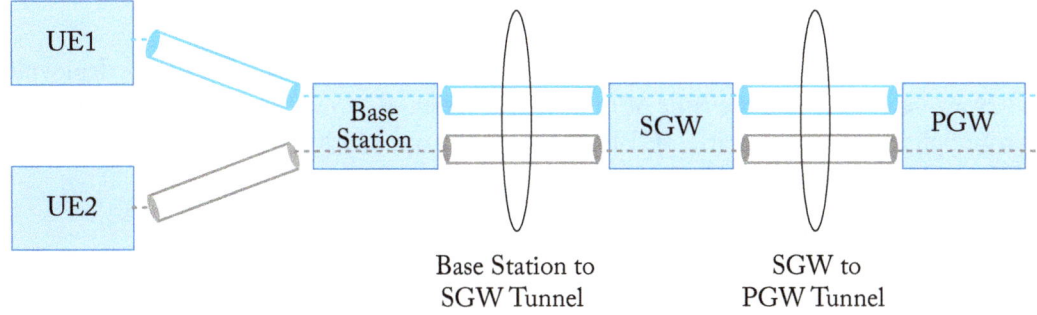

Figure 3.12: Sequence of per-hop tunnels involved in an end-to-end User Plane channel.

the *Advanced Encryption Standard* (AES), where the actual protocol used is an implementation choice. Note that this authentication exchange is in the clear since the Base Station to UE link is not yet secure.

Once the UE and Core-CP are satisfied with each other's identity, the Core-CP informs the other components of the parameters they will need to service the UE (Step 3). This includes: (a) instructing the Core-UP to initialize the user plane (e.g., assign an IP address to the UE and set the appropriate QCI parameter); (b) instructing the Base Station to establish an encrypted channel to the UE; and (c) giving the UE the symmetric key it will need to use the encrypted channel with the Base Station. Once complete, the UE can use the end-to-end user plane channel through the Core-UP (Step 4).

There are three additional details of note about this process. First, the secure control channel between the UE and the Core-CP set up during Step 2 remains available, and is used by the Core-CP to send additional control instructions to the UE during the course of the session.

Second, the user plane channel established during Step 4 is referred to as the *Default Bearer Service*, but additional channels can be established between the UE and Core-UP, each with a potentially different QCI value. This might be done on an application-by-application basis, for example, under the control of the Mobile Core doing *Deep Packet Inspection* (DPI) on the traffic, looking for flows that require special treatment.

Third, while the resulting user plane channels are logically end-to-end, each is actually implemented as a sequence of per-hop tunnels, as illustrated in Figure 3.12. (The figure shows the SGW and PGW from the 4G Mobile Core to make the example more concrete.) This means each component on the end-to-end path terminates a downstream tunnel using one local identifier for a given UE, and initiates an upstream tunnel using a second local identifier for that UE. In practice, these per-flow tunnels are often bundled into an single inter-component tunnel, which makes it impossible to differentiate the level of service given to any particular end-to-end UE channel. This is a limitation of 4G that 5G has ambitions to correct.

3.5   DEPLOYMENT OPTIONS

With an already deployed 4G RAN/EPC in the field and a new 5G RAN/NG-Core deployment underway, we can't ignore the issue of transitioning from 4G to 5G (an issue the IP-world has been grappling with for 20 years). 3GPP officially spells out multiple deployment options, which can be summarized as follows.

- Standalone 4G/Stand-Alone 5G.

- Non-Standalone (4G + 5G RAN) over 4G's EPC.

- Non-Standalone (4G+5G RAN) over 5G's NG-Core.

The second of the three options, which is generally referred to by its NSA acronym, involves 5G base stations being deployed alongside the existing 4G base stations in a given geography to provide a data-rate and capacity boost. In NSA, control plane traffic between the user equipment and the 4G Mobile Core utilizes (i.e., is forwarded through) 4G base stations, and the 5G base stations are used only to carry user traffic. Eventually, it is expected that operators complete their migration to 5G by deploying NG Core and connecting their 5G base stations to it for Standalone (SA) operation. NSA and SA operations are illustrated in Figure 3.13.

One reason we call attention to the phasing issue is that we face a similar challenge in the chapters that follow. The closer the following discussion gets to implementation details, the more specific we have to be about whether we are using 4G components or 5G components. As a general rule, we use 4G components—particularly with respect to the Mobile Core, since that's what's available in open source today—and trust the reader can make the appropriate substitution without loss of generality. Like the broader industry, the open source community is in the process of incrementally evolving its 4G code base into its 5G-compliant counterpart.

Further Reading

For more insight into 4G to 5G migration strategies, see Road to 5G: Introduction and Migration. GSMA Report, April 2018.

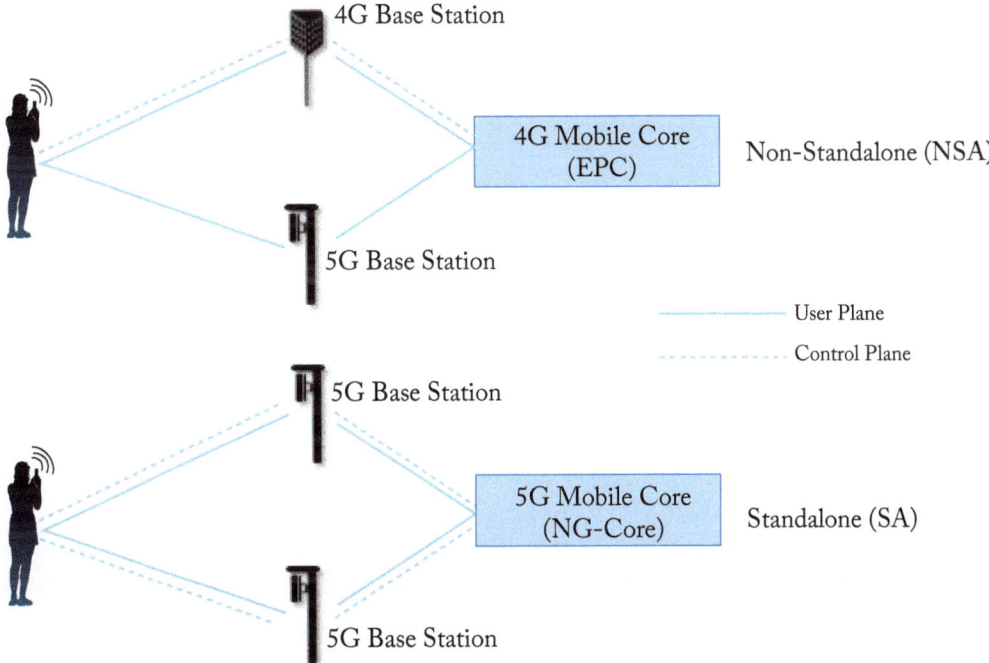

Figure 3.13: NSA and SA options for 5G deployment.

CHAPTER 4

RAN Internals

The description of the RAN in the previous chapter focused on functionality, but was mostly silent about the RAN's internals structure. We now focus in on some of the internal details, and in doing so, explain how the RAN is being transformed in 5G. This involves first describing the stages in the packet processing pipeline, and then showing how these stages can be disaggregated, distributed and implemented.

Our approach in this chapter is to incrementally build the RAN from the bottom up in the first three sections. Section 4.4 then summarizes the overall design, with a focus on how the resulting end-to-end system is architected to evolve.

4.1 PACKET PROCESSING PIPELINE

Figure 4.1 shows the packet processing stages implemented by the base station. These stages are specified by the 3GPP standard. Note that the figure depicts the base station as a pipeline (running left-to-right) but it is equally valid to view it as a protocol stack (as is typically done in official 3GPP documents). Also note that (for now) we are agnostic as to how these stages are implemented, but since we are ultimately heading towards a cloud-based implementation, you can think of each as corresponding to a microservice (if that is helpful).

The key stages are as follows.

- RRC (Radio Resource Control) → Responsible for configuring the coarse-grain and policy-related aspects of the pipeline. The RRC runs in the RAN's control plane; it does not process packets on the user plane.

- PDCP (Packet Data Convergence Protocol) → Responsible for compressing and decompressing IP headers, ciphering and integrity protection, and making an "early" forwarding decision (i.e., whether to send the packet down the pipeline to the UE or forward it to another base station).

- RLC (Radio Link Control) → Responsible for segmentation and reassembly, including reliably transmitting/receiving segments by implementing ARQ.

- MAC (Media Access Control) → Responsible for buffering, multiplexing and demultiplexing segments, including all real-time scheduling decisions about what segments are transmitted when. Also able to make a "late" forwarding decision (i.e., to alternative carrier frequencies, including Wi-Fi).

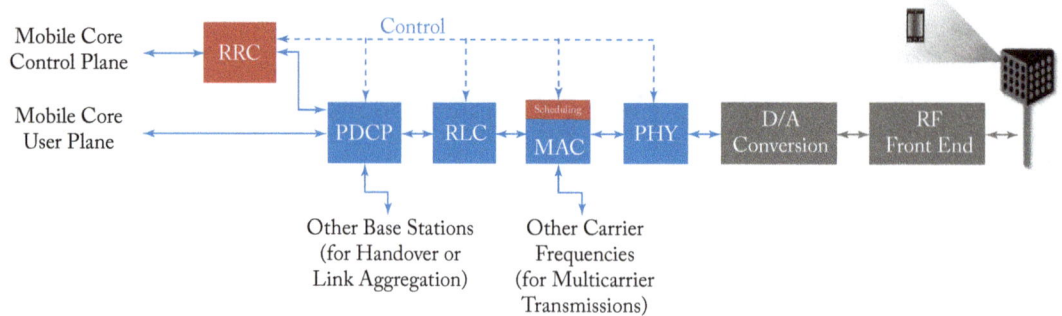

Figure 4.1: RAN processing pipeline, including both user and control plane components.

- PHY (Physical Layer) → Responsible for coding and modulation (as discussed in an earlier chapter), including FEC.

The last two stages in Figure 4.1 (D/A conversion and the RF front-end) are beyond the scope of this book.

While it is simplest to view the stages in Figure 4.1 as a pure left-to-right pipeline, in practice the Scheduler running in the MAC stage implements the "main loop" for outbound traffic, reading data from the upstream RLC and scheduling transmissions to the downstream PHY. In particular, since the Scheduler determines the number of bytes to transmit to a given UE during each time period (based on all the factors outlined in an earlier chapter), it must request (get) a segment of that length from the upstream queue. In practice, the size of the segment that can be transmitted on behalf of a single UE during a single scheduling interval can range from a few bytes to an entire IP packet.

4.2 SPLIT RAN

The next step is to understand how the functionality outlined above is partitioned between physical elements, and hence, "split" across centralized and distributed locations. The dominant option has historically been "no split," with the entire pipeline shown in Figure 4.1 running in the base station. Going forward, the 3GPP standard has been extended to allow for multiple split-points, with the partition shown in Figure 4.2 being actively pursued by the operator-led O-RAN (Open RAN) Alliance. It is the split we adopt throughout the rest of this book.

This results in a RAN-wide configuration similar to that shown in Figure 4.3, where a single *Central Unit (CU)* running in the cloud serves multiple *Distributed Units (DUs)*, each of which in turn serves multiple *Radio Units (RUs)*. Critically, the RRC (centralized in the CU) is responsible for only near-real time configuration and control decision making, while the Scheduler that is part of the MAC stage is responsible for all real-time scheduling decisions.

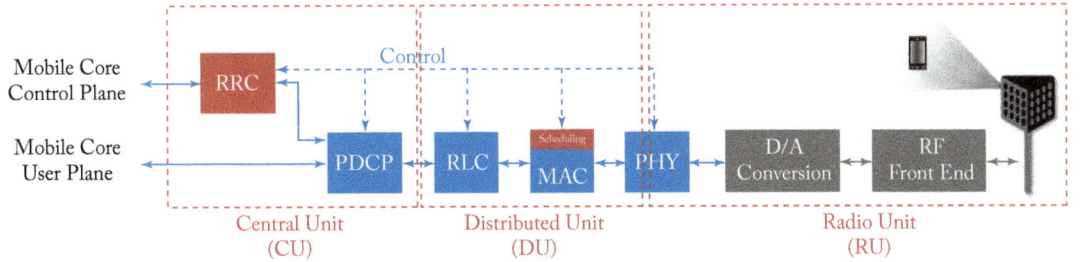

Figure 4.2: Split-RAN processing pipeline distributed across a Central Unit (CU), Distributed Unit (DU), and Radio Unit (RU).

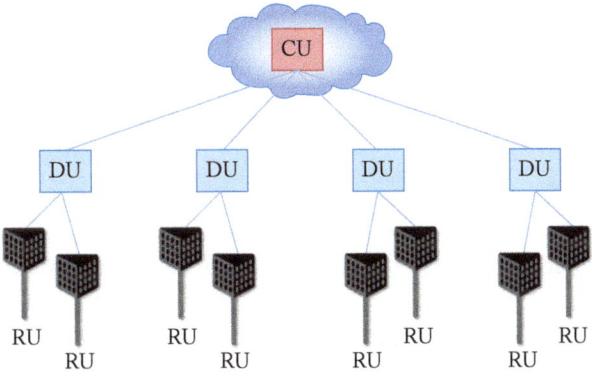

Figure 4.3: Split-RAN hierarchy, with one CU serving multiple DUs, each of which serves multiple RUs.

Clearly, a DU needs to be "near" (within 1 ms) the RUs it manages since the MAC schedules the radio in real-time. One familiar configuration is to co-locate a DU and an RU in a cell tower. But when an RU corresponds to a small cell, many of which might be spread across a modestly sized geographic area (e.g., a mall, campus, or factory), then a single DU would likely service multiple RUs. The use of mmWave in 5G is likely to make this later configuration all the more common.

Also note that the split-RAN changes the nature of the Backhaul Network, which in 4G connected the base stations (eNBs) back to the Mobile Core. With the split-RAN there are multiple connections, which are officially labeled as follows.

- RU-DU connectivity is called the Fronthaul.

- DU-CU connectivity is called the Midhaul.

- CU-Mobile Core connectivity is called the Backhaul.

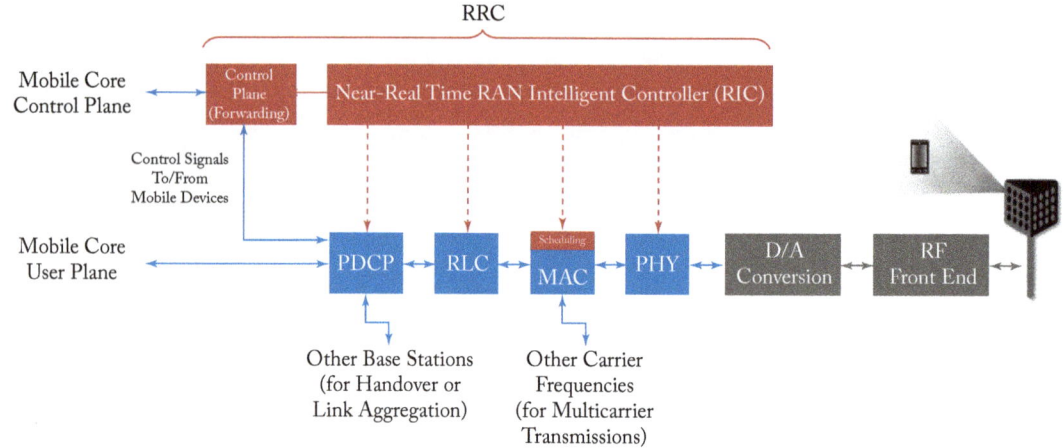

Figure 4.4: RRC disaggregated into a Mobile Core facing control plane component and a Near Real-Time Controller.

One observation about the CU (which is relevant in the next chapter) is that one might co-locate the CU and Mobile Core in the same cluster, meaning the backhaul is implemented in the cluster switching fabric. In such a configuration, the midhaul then effectively serves the same purpose as the original backhaul, and the fronthaul is constrained by the predictable/low-latency requirements of the MAC stage's real-time scheduler.

A second observation about the CU shown in Figure 4.2 is that it encompasses two functional blocks—the RRM and the PDPC—which lie on the RAN's control plane and user plane, respectively. This separation is consistent with the idea of CUPS introduced in Chapter 3, and plays an increasingly important role as we dig deeper into how the RAN is implemented. For now, we note that the two parts are typically referred to as the CU-C and CU-U, respectively.

Further Reading

For more insight into design considerations for interconnecting the distributed components of a Split RAN, see RAN Evolution Project: Backhaul and Fronthaul Evolution. NGMN Alliance Report, March 2015.

4.3   SOFTWARE-DEFINED RAN

We now describe how the RAN is implemented according to SDN principles, resulting in an SD-RAN. The key architectural insight is shown in Figure 4.4, where the RRC from Figure 4.1 is partitioned into two sub-components: the one on the left provides a 3GPP-compliant way

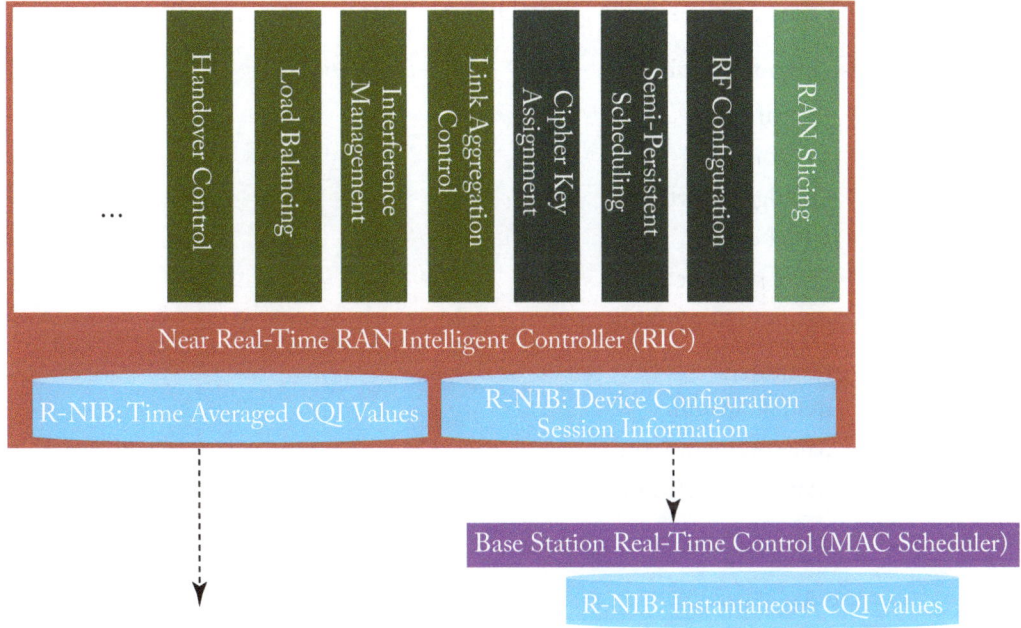

Figure 4.5: Example set of control applications running on top of Near Real-Time RAN Controller.

for the RAN to interface to the Mobile Core's control plane, while the one on the right opens a new programmatic API for exerting software-based control over the pipeline that implements the RAN user plane.

To be more specific, the left sub-component simply forwards control packets between the Mobile Core and the PDCP, providing a path over which the Mobile Core can communicate with the UE for control purposes, whereas the right sub-component implements the core of the RCC's control functionality. This component is commonly referred to as the *RAN Intelligent Controller (RIC)* in O-RAN architecture documents, so we adopt this terminology. The "Near-Real Time" qualifier indicates the RIC is part of 10–100 ms control loop implemented in the CU, as opposed to the ~1 ms control loop required by the MAC scheduler running in the DU.

Although not shown in Figure 4.4, keep in mind (from Figure 4.2) that all constituent parts of the RRC, plus the PDCP, form the CU.

Completing the picture, Figure 4.5 shows the Near-RT RIC implemented as a traditional SDN Controller hosting a set of SDN control apps. The RIC maintains a *RAN Network Information Base (R-NIB)* that includes time-averaged CQI values and other per-session state (e.g., GTP tunnel IDs, QCI values for the type of traffic), while the MAC (as part of the DU) maintains the instantaneous CQI values required by the real-time scheduler. Specifically, the R-NIB includes the following state.

- NODES: Base Stations and Mobile Devices.

  - Base Station Attributes:
    * Identifiers
    * Version
    * Config Report
    * RRM config
    * PHY resource usage

  - Mobile Device Attributes:
    * Identifiers
      item Capability
    * Measurement Config
    * State (Active/Idle)

- LINKS: *Actual* between two nodes and *Potential* between UEs and all neighbor cells.

  - Link Attributes:
    * Identifiers
    * Link Type
    * Config/Bearer Parameters
    * QCI Value

- SLICES: Virtualized RAN Construct.

  - Slice Attributes:
    * Links
    * Bearers/Flows
    * Validity Period
    * Desired KPIs
    * MAC RRM Configuration
    * RRM Control Configuration

The example Control Apps in Figure 4.5 include a range of possibilities, but is not intended to be an exhaustive list. The right-most example, RAN Slicing, is the most ambitious in that it introduces a new capability: Virtualizing the RAN. It is also an idea that has been implemented, which we describe in more detail in the next chapter.

The next three (RF Configuration, Semi-Persistent Scheduling, Cipher Key Assignment) are examples of configuration-oriented applications. They provide a programmatic way to manage seldom-changing configuration state, thereby enabling zero-touch operations. Coming up with meaningful policies (perhaps driven by analytics) is likely to be an avenue for innovation in the future.

The left-most four example Control Applications are the sweet spot for SDN. These functions—Link Aggregation Control, Interference Management, Load Balancing, and Handover Control—are currently implemented by individual base stations with only local visibility, but they have global consequenes. The SDN approach is to collect the available input data centrally, make a globally optimal decision, and then push the respective control parametes back to the base stations for execution. Realizing this value in the RAN is still a work-in-progress, but evidence using the same approach to optimize wide-area networks is compelling.

While the above loosely categorizes the space of potential control apps as either config-oriented or control-oriented, another possible characterization is based on the current practice of controlling the mobile link at two different levels. At a fine-grain level, per-node and per-link control is conducted using Radio Resource Management (RRM) functions that are distributed across the individual base stations. RRM functions include scheduling, handover control, link and carrier aggregation control, bearer control, and access control. At a coarse-grain level, regional mobile network optimization and configuration is conducted using *Self-Organizing Network (SON)* functions. These functions oversee neighbor lists, manage load balancing, optimize coverage and capacity, aim for network-wide interference mitigation, centrally configure parameters, and so on. As a consequence of these two levels of control, it is not uncommon to see reference to *RRM Applications* and *SON Applications*, respectively, in O-RAN documents for SD-RAN.

Further Reading

For an example of how SDN principles have been successfully applied to a production network, we recommend B4: Experience with a Globally-Deployed Software Defined WAN. ACM SICOMM, August 2013.

4.4 ARCHITECT TO EVOLVE

We conclude this description of RAN internals by re-visiting the sequence of steps involved in disaggregation, which as the previous three sections reveal, is being pursued in multiple tiers. In doing so, we tie up several loose ends, including the new interfaces disaggregation exposes. These interfaces define the pivot points around which 5G RAN is architected to evolve.

In the first tier of disaggregation, 3GPP standards provide multiple options of how horizontal RAN splits can take place. Horizontal disaggregation basically splits the RAN pipeline shown in Figure 4.1 into independently operating components. Figure 4.6a illustrates horizontal disaggregation of the RAN from a single base station into three distinct components: CU, DU and RU. The O-RAN Alliance has selected specific disaggregation options from 3GPP and is

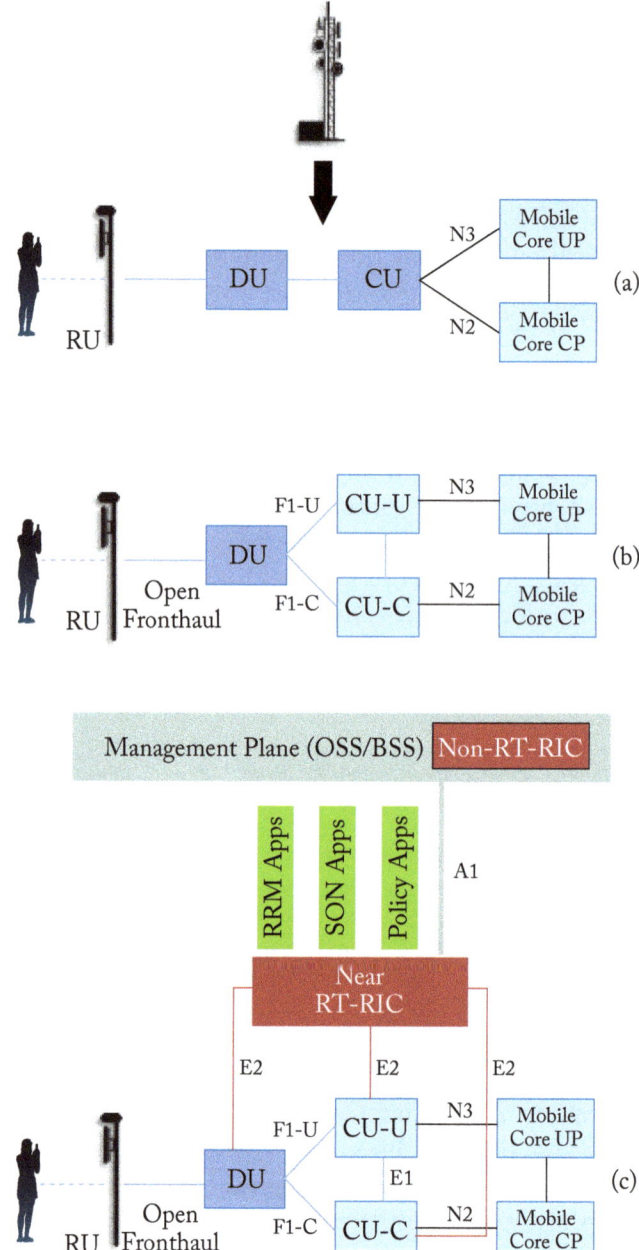

Figure 4.6: Three tiers of RAN disaggregation: (a) horizontal, (b) vertical CUPS, and (c) vertical SDN.

developing open interfaces between these components. 3GPP defines the **N2** and **N3** interfaces between the RAN and the Mobile Core.

The second tier of disaggregation is vertical, focusing on control/user plane separation (CUPS) of the CU, and resulting in CU-U and CU-C shown in Figure 4.6b. The control plane in question is the 3GPP control plane, where the CU-U realizes a pipeline for user traffic and the CU-C focuses on control message signaling between Mobile Core and the disaggregated RAN components (as well as to the UE). The O-RAN specified interfaces between these disaggregated components are also shown in Figure 4.6b.

The third tier follows the SDN paradigm by carrying vertical disaggregation one step further. It does this by separating most of RAN control (RRM functions) from the disaggregated RAN components, and logically centralizing them as applications running on an SDN Controller, which corresponds to the Near-RT RIC shown previously in Figures 4.4 and 4.5. This SDN-based vertical disaggregation is repeated here in Figure 4.6c. The figure also shows the additional O-RAN prescribed interfaces.

The interface names are cryptic, and knowing their details adds little to our conceptual understanding of the RAN, except perhaps to re-enforce how challenging it is to introduce a transformative technology like Software-Defined Networking into an operational environment that is striving to achieve full backward compatibility and universal interoperability. That said, we call out two notable examples.

The first is the **A1** interface that the mobile operator's management plane—typically called the *OSS/BSS (Operations Support System/Business Support System)* in the Telco world—uses the to configure the RAN. We have not discussed the Telco OSS/BSS up to this point, but it safe to assume such a component sits at the top of any Telco software stack. It is the source of all configuration settings and business logic needed to operate a network. Notice that the Management Plane shown in Figure 4.6c includes a *Non–Real TIme RIC* functional block, complementing the Near-RT RIC that sits below the A1 interface. We return to the relevance of these two RICs in a moment.

The second is the **E2** interface that the Near-RT RIC uses to control the underlying RAN elements. A requirement of the E2 interface is that it be able to connect the Near-RT RIC to different types of RAN elements. This range is reflected in the API, which revolves around a *Service Model* abstraction. The idea is that each RAN element advertises a Service Model, which effectively defines the set of RAN Functions the element is able to support. The RIC then issues a combination of the following four operations against this Service Model.

- **Report:** RIC asks the element to report a function-specific value setting.

- **Insert:** RIC instructs the element to activate a user plane function.

- **Control:** RIC instructs the element to activate a control plane function.

- **Policy:** RIC sets a policy parameter on one of the activated functions.

Of course, it is the RAN element, through its published Service Model, that defines the relevant set of functions that can be activated, the variables that can be reported, and policies that can be set.

Taken together, the A1 and E2 interfaces complete two of the three major control loops of the RAN: the outer (non-realtime) loop has the Non-RT RIC as it control point and the middle (near-realtime) loop has the Near-RT RIC as its control point. The third (inner) control loop, which is not shown in Figure 4.6, runs inside the DU: it includes the realtime Scheduler embedded in the MAC stage of the RAN pipeline. The two outer control loops have rough time bounds of >>1 sec and >10 ms, respectively, and as we saw in Chapter 2, the realtime control loop is assumed by be <1 ms.

This raises the question of how specific functionality is distributed between the Non-RT RIC, Near-RT RIC, and DU. Starting with the second pair (i.e., the two inner loops), it is important to recognize that not all RRM functions can be centralized. After horizontal and vertical CUPS disaggregation, the RRM functions are split between CU-C and DU. For this reason, the SDN-based vertical disaggregation focuses on centralizing CU-C-side RRM functions in the Near-RT RIC. In addition to RRM control, this includes all the SON applications.

Turning to the outer two control loops, the Near RT-RIC opens the possibility of introducing policy-based RAN control, whereby interrupts (exceptions) to operator-defined policies would signal the need for the outer loop to become involved. For example, one can imagine developing learning-based controls, where the inference engines for these controls would run as applications on the Near RT-RIC, and their non-realtime learning counterparts would run elsewhere. The Non-RT RIC would then interact with the Near-RT RIC to deliver relevant operator policies from the Management Plane to the Near RT-RIC over the A1 interface.

Finally, you may be wondering why there is an O-RAN Alliance in the first place, given that 3GPP is already the standardization body responsible for interoperability across the global cellular network. The answer is that over time 3GPP has become a vendor-dominated organization, whereas O-RAN was created more recently by network operators. (AT&T and China Mobile were the founding members.) O-RAN's goal is to catalyze a software-based implementation that breaks the vendor lock-in that dominates today's marketplace. The E2 interface in particular, which is architected around the idea of supporting different Service Models, is central to this strategy. Whether the operators will be successful in their ultimate goal is yet to be seen.

CHAPTER 5

Advanced Capabilities

Disaggregating the cellular network pays dividends. This chapter explores three examples. Stepping back to look at the big picture, Chapter 3 (Architecture) described "what is" (the basics of 3GPP) and Chapter 4 (RAN Internals) described "what will be" (where the industry is clearly headed), whereas this chapter describes "what might be" (our best judgement on cutting-edge capabilities that will eventually be realized).

5.1 OPTIMIZED DATA PLANE

There are many reasons to disaggregate functionality, but one of the most compelling is that by decoupling control and data code paths, it is possible to optimize the data path. This can be done, for example, by programming it into specialized hardware. Modern white-box switches with programmable packet forwarding pipelines are a perfect example of specialized hardware we can exploit in the cellular network. Figure 5.1 shows the first step in the process of doing this. The figure also pulls together all the elements we've been describing up to this point. There are several things to note about this diagram.

First, the figure combines both the Mobile Core and RAN elements, organized according to the major subsystems: Mobile Core, Central Unit (CU), Distributed Unit (DU), and Radio Unit (RU). The figure also shows one possible mapping of these subsystems onto physical locations, with the first two co-located in a Central Office and the latter two co-located in a cell tower. As discussed earlier, other configurations are also possible.

Second, the figure shows the Mobile Core's two user plane elements (PGW, SGW) and the Central Unit's single user plane element (PDCP) further disaggregated into control/user plane pairs, denoted PGW-C/PGW-U, SGW-C/SGW-U, and PDCP-C/PDCP-U, respectively. Exactly how this decoupling is realized is an implementation choice (i.e., not specified by 3GPP), but the idea is to reduce the User Plane component to the minimal Receive-Packet/Process-Packet/Send-Packet processing core, and elevate all control-related aspects into the Control Plane component.

Third, the PHY (Physical) element of the RAN pipeline is split between the DU and RU partition. Although beyond the scope of this book, the 3GPP spec specifies the PHY element as a collection of functional blocks, some of which can be effectively implemented by software running on a general-purpose processor and some of which are best implemented in specialized hardware (e.g., a Digital Signal Processor). These two subsets of functional blocks map to the PHY Upper (part of the DU) and the PHY Lower (part of the RU), respectively.

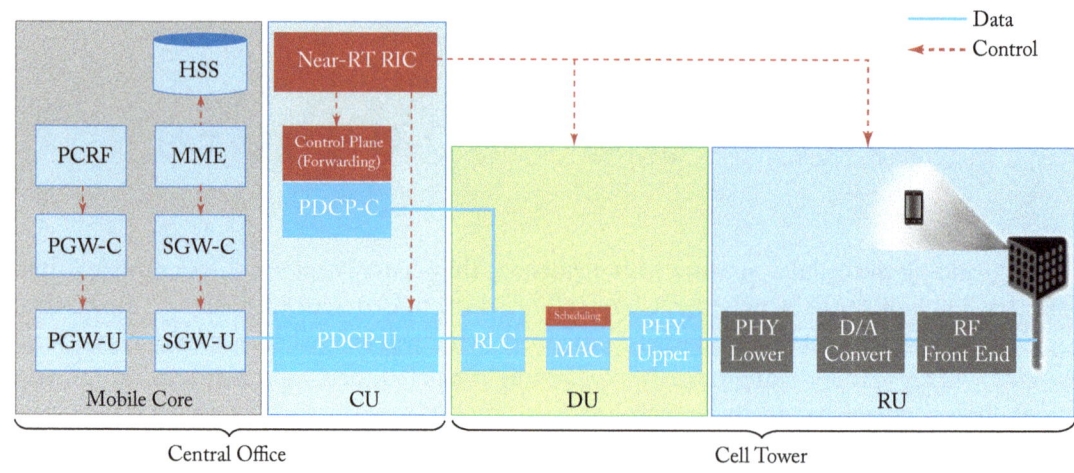

Figure 5.1: End-to-end disaggregated system, including Mobile Core and Split-RAN.

Fourth, and somewhat confusingly, Figure 5.1 shows the PCDP-C element and the Control Plane (Forwarding) element combined into a single functional block, with a data path (blue line) connecting that block to both the RLC and the MME. Exactly how this pair is realized is an implementation choice (e.g., they could map onto two or more microservices), but the end result is that they are part of an end-to-end path over which the MME can send control packets to the UE. Note that this means responsibility for demultiplexing incoming packets between the control plane and user plane falls to the RLC.

Figure 5.2 shows why we disaggregated these components: it allows us to realize the three user plane elements (PGW-U, SGW-U, PDCP-U) in switching hardware. This can be done using a combination of a language that is tailored for programming forwarding pipelines (e.g., P4), and a protocol-independent switching architecture (e.g., Tofino). For now, the important takeaway is that the RAN and Mobile Core user plane can be mapped directly onto an SDN-enabled data plane.

Further Reading

For more information about P4 and programmable switching chips, we recommend White-Box Switches, a chapter in *Software-Defined Networking: A Systems Approach*, March 2020.

Pushing RAN and Mobile Core forwarding functionality into the switching hardware results in overlapping terminology that can be confusing. 5G separates the functional blocks into control and user planes, while SDN disaggregates a given functional block into control and data plane halves. The overlap comes from our choosing to implement the 5G user plane

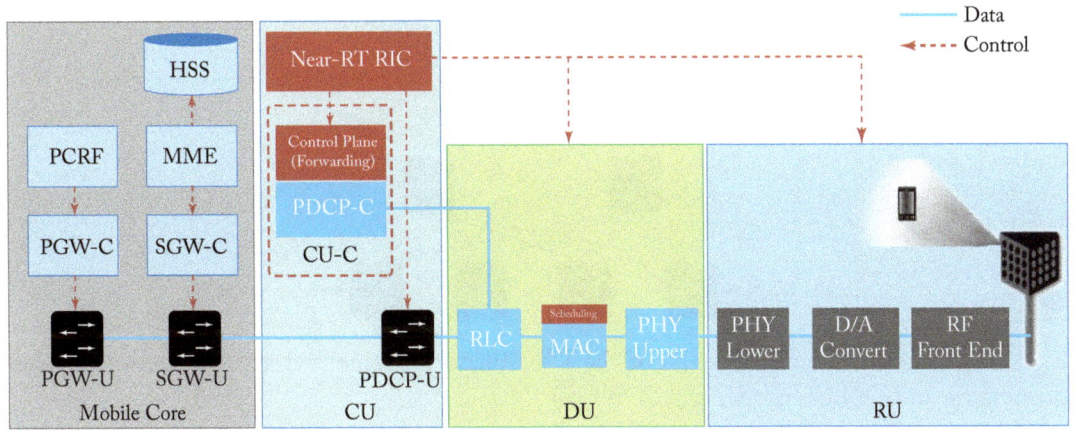

Figure 5.2: Implementing data plane elements of the User Plane in programmable switches.

by splitting it into its SDN-based control and data plane parts. As one simplification, we refer to the Control Plane (Forwarding) and PDCP-C combination as the CU-C (Central Unit – Control) going forward.

Finally, the SDN-prescribed control/data plane disaggregation comes with an implied implementation strategy, namely, the use of a scalable and highly available *Network Operating System (NOS)*. Like a traditional OS, a NOS sits between application programs (control apps) and the underlying hardware devices (whitebox switches), providing higher levels of abstraction (e.g., network graph) to those applications, while hiding the low-level details of the underlying hardware. To make the discussion more concrete, we use ONOS (Open Network Operating System) as an example NOS, where PGW-C, SGW-C, and PDCP-C are all realized as control applications running on top of ONOS.

Figure 5.3 shows one possible configuration in which the underlying switches are interconnected to form a leaf-spine fabric. Keep in mind that the linear sequence of switches implied by Figure 5.2 is logical, but that in no way restricts the actual hardware to the same topology. The reason we use a leaf-spine topology is related to our ultimate goal of building an edge cloud, and leaf-spine is the proto-typical structure for such cloud-based clusters. This means the three control applications must work in concert to implement an end-to-end path through the fabric, which in practice happens with the aid of other, fabric aware, control applications (as implied by the "..." in the Figure). We describe the complete picture in more detail in the next chapter, but for now, the big takeaway is that the control plane components of the 5G overlay can be realized as control applications for an SDN-based underlay.

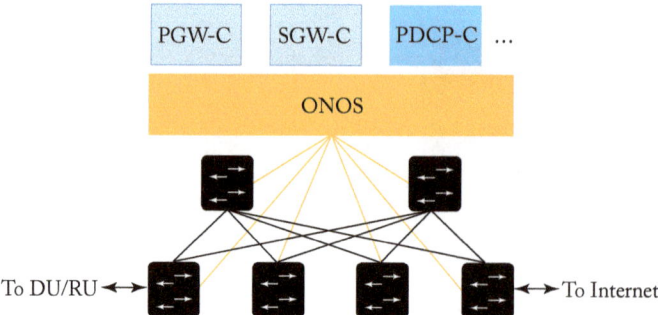

Figure 5.3: Control Plane elements of the User Plane implemented as Control Applications running on an SDN Controller (e.g., ONOS).

5.2   MULTI-CLOUD

Another consequence of disaggregating functionality is that once decoupled, different functions can be placed in different physical locations. We have already seen this when we split the RAN, placing some functions (e.g., the PCDP and RRC) in the Central Unit and others (e.g., RLC and MAC) in Distributed Units. This allows for simpler (less expensive) hardware in remote locations, where there are often space, power, and cooling constraints.

This process can be repeated by distributing the more centralized elements across multiple clouds, including large datacenters that already benefit from elasticity and economies of scale. Figure 5.4 shows the resulting multi-cloud realization of the Mobile Core. We leave the user plane at the edge of the network (e.g., in the Central Office) and move control plane to a centralized cloud. It could even be a public cloud like Google or Amazon. This includes not only the MME, PCRF, and HSS, but also the PGW-C and SGW-C we decoupled in the previous section. (Note that Figure 5.4 renames the PDCP-U from earlier diagrams as the CU-U; either label is valid.)

What is the value in doing this? Just like the DU and RU, the Edge Cloud likely has limited resources. If we want room to run new edge services there, it helps to move any components that need not be local to a larger facility with more abundant resources. Centralization also facilitates collecting and analyzing data across multiple edge locations, which is harder to do if that information is distributed over multiple sites. (Analytics performed on this data also benefits from having abundant compute resources available.)

Another reason worth calling out is that it lowers the barrier for anyone (not just the companies that own and operate the RAN infrastructure) to offer mobile services to customers. These entities are called *MVNOs (Mobile Virtual Network Operators)* and one clean way to engineer an MVNO is to run your own Mobile Core in a cloud of your choosing.

But the biggest motivation for the configuration shown in Figure 5.4 is that keeping the user plane elements of the mobile core at the edge makes it possible to "break out" local traffic

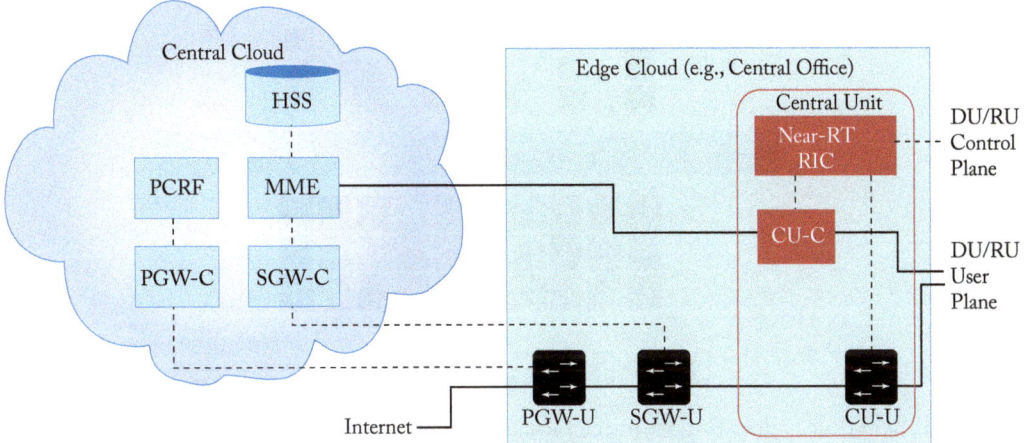

Figure 5.4: Multi-Cloud implementation, with MME, HSS, PCRF, and Control Plane elements of the PGW and SGW running in a centralized cloud.

without having to traverse a "turnpin" route that goes through a central site. This has the potential to dramatically reduce latency for any edge-hosted services. We return to this topic in Chapter 7.

5.3   NETWORK SLICING

One of the most compelling value propositions of 5G is the ability to differentiate the level of service offered to different applications and customers. Differentiation, of course, is key to being able to charge some customers more than others, but the monetization case aside, it is also necessary if you are going to support such widely varying applications as streaming video (which requires high bandwidth but can tolerate larger latencies) and IoT (which has minimal bandwidth needs but requires extremely low and predictable latencies, connecting a *massively scalable* number of IoT devices).

The mechanism that supports this sort of differentiation is called network slicing, and it fundamentally comes down to scheduling, both in the RAN (deciding which segments to transmit) and in the Mobile Core (scaling microservice instances and placing those instances on the available servers). The following introduces the basic idea, starting with the RAN.

But before getting into the details, we note that a network slice is a realization of the QoS Class Index (QCI) discussed earlier. 3GPP specifies a standard set of network slices, called *Standardized Slice Type (SST)* values. For example, SST 1 corresponds to mobile broadband, SST 2 corresponds to Ultra-Reliable Low Latency Communications, SST 3 corresponds to Massive IoT, and so on. It is also possible to extend this standard set with additional slice behaviors, as well as define multiple slices for each SST (e.g., to further differentiate subscribers based on priority).

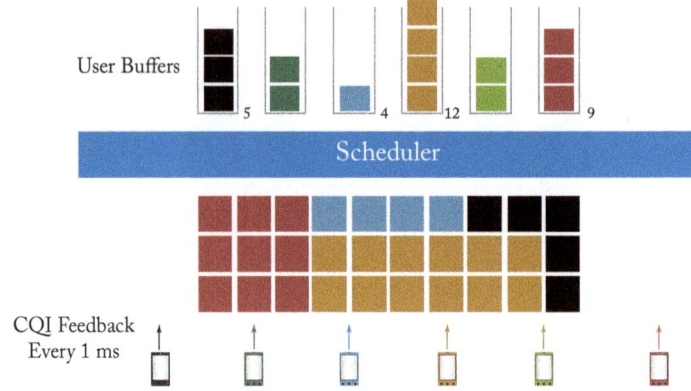

Figure 5.5: Scheduler allocating resource blocks to UEs.

Of course, defining a desired set of slices and implementing a slicing mechanism are two entirely different things. The following sketches how slices can be implemented.

5.3.1   RAN SLICING

We start by reviewing the basic scheduling challenge previewed in Chapter 2. As depicted in Figure 5.5, the radio spectrum can be conceptualized as a 2-D grid of *Resource Blocks (RB)*, where the scheduler's job is to decide how to fill the grid with the available segments from each user's transmission queue based on CQI feedback from the UEs. To restate, the power of OFDMA is the flexibility it provides in how this mapping is performed.

While in principle one could define an uber scheduler that takes dozens of different factors into account, the key to network slicing is to add a layer of indirection. As shown in Figure 5.6, the idea is to perform a second mapping of Virtual RBs to Physical RBs. This sort of virtualization is common in resource allocators throughout computing systems because we want to separate how many resources are allocated to each user from the decision as to which physical resources are actually assigned. This virtual-to-physical mapping is performed by a layer typically known as a *Hypervisor*, and the important thing to keep in mind is that it is totally agnostic about which user's segment is affected by each translation.

Having decoupled the Virtual RBs from Physical RBs, it is now possible to define multiple Virtual RB sets (of varying sizes), each with its own scheduler. Figure 5.7 gives an example with two equal-sized RB sets, where the important consequence is that having made the macro-decision that the Physical RBs are divided into two equal partitions, the scheduler associated with each partition is free to allocate Virtual RBs completely independent from each other. For example, one scheduler might be designed to deal with high-bandwidth video traffic and another scheduler might be optimized for low-latency IoT traffic. Alternatively, a certain fraction of the

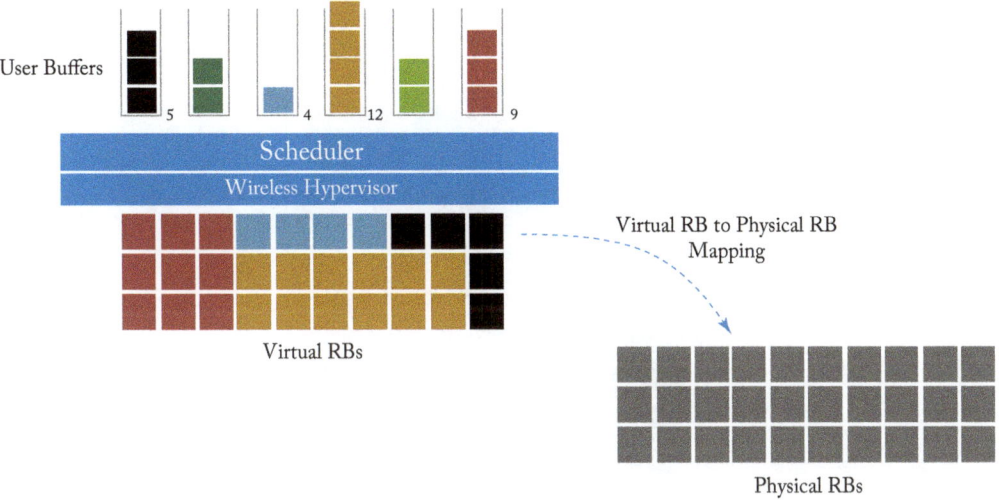

Figure 5.6: Wireless Hypervisor mapping virtual resource blocks to physical resource blocks.

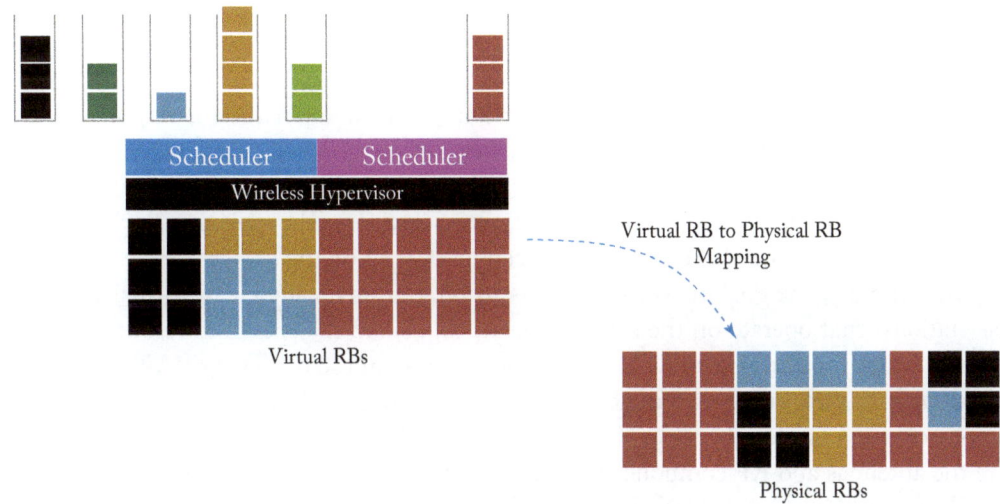

Figure 5.7: Multiple schedulers running on top of wireless hypervisor.

available capacity could be reserved for premium customers or other high-priority traffic (e.g., public safety), with the rest shared among everyone else.

Going one level deeper in the implementation details, the real-time scheduler running in each DU receives high-level directives from the near real-time scheduler running in the CU, and as depicted in Figure 5.8, these directives make dual transmission, handoff, and interference

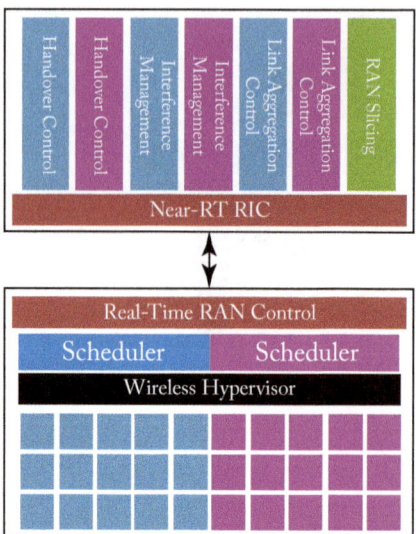

Figure 5.8: Centralized near-realtime control applications cooperating with distribute real-time RAN schedulers.

decisions on a per-slice basis. A single RAN Slicing control application is responsible for the macro-scheduling decision by allocating resources among a set of slices. Understanding this implementation detail is important because all of these control decisions are implemented by software modules, and hence, easily changed or customized. They are not "locked" into the underlying system, as they have historically been in 4G's eNodeBs.

In summary, the goal of RAN slicing is to programmatically create virtual RAN nodes (base stations) that operate on the same hardware and share the spectrum resources according to a given policy for different applications, services, users, and so on. Tying RAN slicing back to RAN disaggregation, one can imagine several possible configurations, depending on the desired level of isolation between the slices. Figure 5.9 shows four examples, all of which assume slices share the antennas and RF components, which is effectively the RU: (a) RAN slices share RU, DU, CU-U, and CU-C; (b) RAN slices share RU and DU, but have their own CU-U and CU-C; (c) RAN slices share RU, CU-U and CU-C, but have their own DU; and (d) RAN slices share RU, but have their own DU, CU-U, and CU-C.

5.3.2   CORE SLICING

In addition to slicing the RAN, we also need to slice the Mobile Core. In many ways, this is a well-understood problem, involving QoS mechanisms in the network switches (i.e., making sure packets flow through the switching fabric according to the bandwidth allocated to each slice)

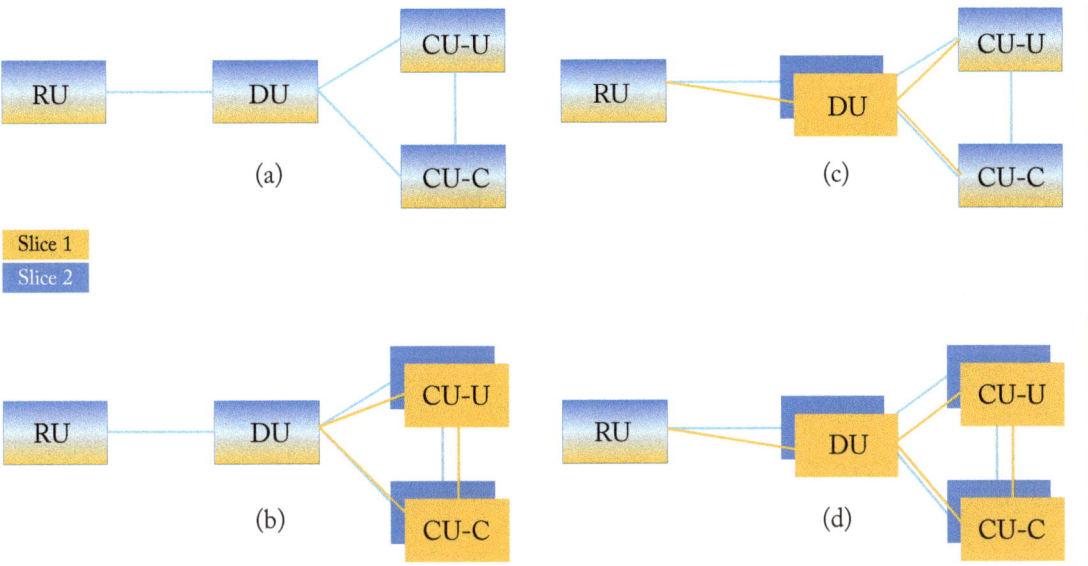

Figure 5.9: Four possible configurations of a disaggregated RAN in support of RAN slicing.

and the cluster processors (i.e., making sure the containers that implement each microservice are allocated sufficient CPU cores to sustain the packet forwarding rate of the corresponding slice).

But packet scheduling and CPU scheduling are low-level mechanisms. What makes slicing work is to also virtualize and replicate the entire service mesh that implements the Mobile Core. If you think of a slice as a system abstraction, then that abstraction needs to keep track of the set of interconnected microservices that implement each slice, and then instruct the underlying packet schedulers to allocate sufficient network bandwidth to the slice's flows and the underlying CPU schedulers to allocate sufficient compute cycles to the slice's containers.

For example, if there are two network slices (analogous to the two RAN schedulers shown in Figures 5.7 and 5.8), then there would also need to be two Mobile Core service meshes: one set of AMF, SMF, UPF,...microservices running on behalf of the first slice and a second set of AMF, SMF, UPF,...microservices running on behalf of the second slice. These two meshes would scale independently (i.e., include a different number of container instances), depending on their respective workloads and QoS guarantees. The two slices would also be free to make different implementation choices, for example, with one optimized for massive IoT applications and the other optimized for high-bandwidth AR/VR applications.

The one remaining mechanism we need is a demultiplexing function that maps a given packet flow (e.g., between UE and some Internet application) onto the appropriate instance of the service mesh. This is the job of the NSSF described in an Chapter 3: it is responsible for selecting the mesh instance a given slice's traffic is to traverse.

We conclude this discussion of slicing with an observation. While differentiating slices based on the resources allocated to each is a familiar network feature, reminiscent of QoS, slices can also implement different functionality, specialized for different use cases. For example, the AMF/SMF (5G) or MME (4G) functionality of the Mobile Core can be customized for different usage patterns, where supporting a massively scalable number of IoT devices that intermittently transmit small amounts of data is a great example. Not only does this break the cellular network out of a one-size-fits-all situation, it opens the door for innovation. There won't be just one Mobile Core. There will potentially be many (and they will be implemented in the cloud).

CHAPTER 6

Exemplar Implementation

The steps we've taken in the previous chapters to virtualize, disaggregate, optimize, distribute, and slice the cellular network not only help us understand the inner-workings of 5G, but they are also necessary to reduce the entirety of the 5G cellular network to practice. The goal is an implementation, which by definition, forces us to make specific engineering choices. This chapter describes one set of engineering choices that results in a running system. It should be interpreted as an exemplar, for the sake of completeness, but not the only possibility.

The system we describe is called CORD, which you will recall from the Introduction is an acronym (**C**entral **O**ffice **R**e-architected as a **D**atacenter). More concretely, CORD is a blueprint for building a 5G deployment from commodity hardware and a collection of open source software components. We call this hardware/software combination a CORD POD, where the idea is to deploy a POD at each edge site that is part of a cellular network. The following describes CORD in terms of a set of engineering decisions. It is not a substitute for detailed documentation for installing, developing, and operating CORD. Also keep in mind that even though CORD includes "Central Office" in its name, a CORD POD is a general design, and not strictly limited to being deployed in a conventional Central Office.

Further Reading

To learn how to install, operate, and contribute to the CORD open source software platform, see the CORD Guide. ONF, March 2020.

Before getting into the specifics, it is important to understand that CORD is a work-in-progress, with a sizable open source community contributing to its code base. Many of the components are quite mature, and currently running in operator trials and production networks. Others (largely corresponding to the advanced capabilities described in the previous chapter) are prototypes that run in "demonstration mode," but are not yet complete enough to be included in official releases. Also, as outlined in the earlier discussion on deployment options, CORD starts with a production-quality EPC that is being incrementally evolved into its 5G counterpart. (This chapter uses the EPC-specific components for illustrative purposes.)

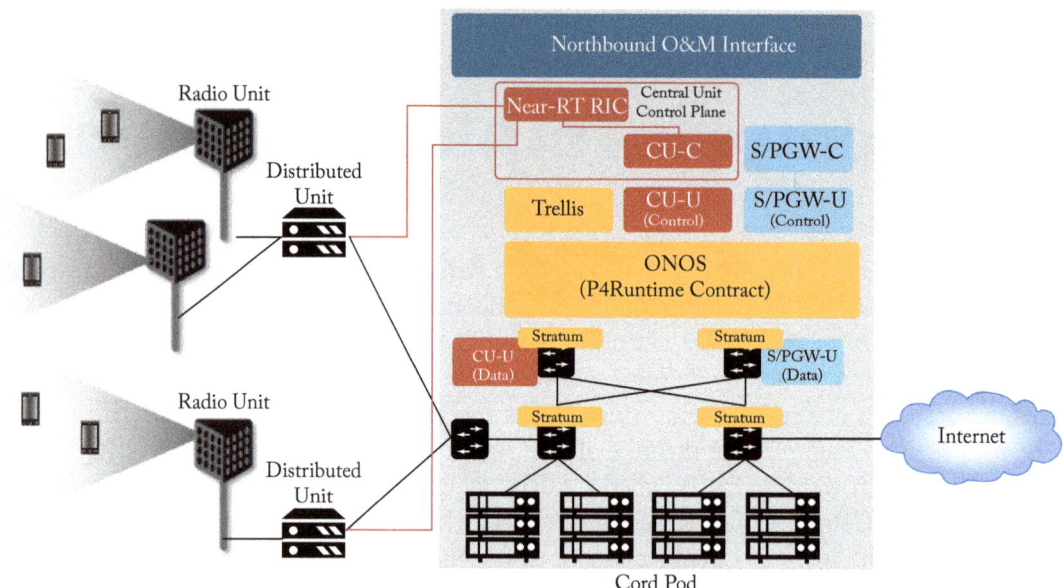

Figure 6.1: CORD implementation of RAN and Mobile Core.

6.1 FRAMEWORK

Figure 6.1 gives a schematic overview of a CORD POD. It connects downstream to a set of DUs (and associated RUs), and upstream to the rest of the Internet. Internally, it includes a set of commodity servers (the figure shows four racks of three servers each, but the design accommodates anywhere from a partial rack to 16 racks) and a set of white-box switches arranged in a leaf-spine topology (the figure shows two leaves and two spine switches, but the design allows anywhere from a single switch to two leaf switches per rack and as many spine switches as necessary to provide sufficient east-to-west bandwidth).

With respect to software, Figure 6.1 shows a combination of RAN (red) and Mobile Core (blue) components, plus the modules that define the CORD platform (orange). We describe the platform components later in this chapter, but you can think of them as collectively implementing a multi-tenant cloud on top of which many different scalable services can run. The RAN and Mobile Core are two such tenants. The CORD platform can also host other edge services (which is one reason CORD is built using cloud technology in the first place), but exactly what other edge services might run on a given CORD POD is a question we do not try to answer in this book.

The RAN and Core related components are ones we've described in earlier chapters. They include the Control and User planes of the CU and Mobile Core, respectively, where to simplify the diagram, we show the SGW and PGW merged into a single S/PGW. One other detail that

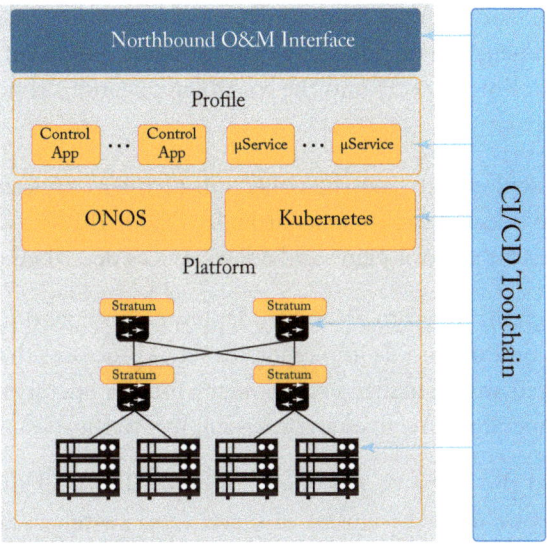

Figure 6.2: Alternative view of CORD, with a CI/CD toolchain managing the platform and set of services implemented by a combination of ONOS-based control apps and Kubernetes-based microservices.

deserves closer attention is the RAN Control component included in the CU Control Plane. This is the Near-RT RIC introduced in Section 4.3, which means a CORD POD includes two SDN Controllers: the RIC controls the RAN and ONOS shown in Figure 6.1 controls the fabric. (The RIC running in CORD actually corresponds to a second, customized version of ONOS, but that's an implementation detail.)

One aspect of Figure 6.1 that requires further elaboration is how each of the RAN and Mobile Core components are actually realized. Specifically, there are three different manifestations of the functional components implied by the figure: (1) the data plane layer of the CU-U and S/PGW-U are realized as P4 programs loaded into the programmable switches; (2) the control plane layer of the CU-U and S/PGW-U (as well as the Trellis platform module) are realized as control applications loaded onto the ONOS Network OS; and (3) the remaining components are realized as Kubernetes-managed microservices. (Kubernetes is implied, but not shown in the figure.)

To expand on this idea, Figure 6.2 gives an alternative view of a CORD POD, abstracting away the details of *what* services it hosts, and focusing instead on *how* those services are instantiated on the POD. In this figure, all the functionality instantiated onto the POD runs as a combination of Kubernetes-based microservices and ONOS-based control applications.

When abstracted in this way, we can view a POD as including three major subsystems.

- **Platform:** The base layer common to all configurations includes Kubernetes as the container management system and ONOS as the SDN controller, with Stratum loaded on to each switch. Both ONOS and the control applications it hosts run as a Kubernetes-managed microservices.

- **Profile:** The deployment-specific collection of microservices and SDN control apps that have been selected to run on a particular POD. This is a variable and evolvable set, and it includes the control plane and edge services described elsewhere.

- **CI/CD Toolchain:** Used to assemble, deploy, operate, and upgrade a particular Platform/Profile combination. It implements a set of processes that transforms a collection of disaggregated and virtualized components into an operational system capable of responding to operator directives and carrying live traffic.

Although beyond the scope of this book, the CI/CD toolchain uses standard DevOps tools to bootstrap software onto the cluster of servers and switches, and rollout/rollback individual microservices and control applications. It also auto-generates the Northbound Interface (NBI) that operators use to manage the POD, based on a declarative specification of the Profile the POD is configured to support. This NBI is sufficiently complete to operate a CORD POD in a production environment.

6.2   PLATFORM COMPONENTS

We now return to the three platform-related components shown in Figures 6.1 and 6.2. Each is a substantial open source component in its own right, but for our purposes, it is enough to understand the role they play in supporting a 5G-based profile of CORD.

- **Stratum:** A thin operating system that runs locally on each white-box switch. Its purpose is to provide a hardware-independent interface for managing and programming the switches in CORD. This includes using *P4* to define the forwarding behavior of the switch's forwarding pipeline (think of this program as a contract between the control and data planes), and *P4Runtime* to control that forwarding contract at runtime.

- **ONOS:** A Network Operating System used to configure and control a network of programmable white-box switches. It runs off-switch as a logically centralized SDN controller, and hosts a collection of SDN control applications, each of which controls some aspect of the underlying network. Because it is logically centralized, ONOS is designed to be highly available and to have scalable performance.

- **Trellis:** An ONOS-hosted SDN control application that implements a leaf-spine fabric on a network of white-box switches. It implements the control plane for several features, including VLANs and L2 bridging, IPv4 and IPv6 unicast and multicast routing, DHCP L3 relay, dual-homing of servers and upstream routers, QinQ forwarding/termination, MPLS pseudowires, and so on. In addition, Trellis can make the entire fabric appear as a single (virtual) router to upstream routers, which communicate with Trellis using standard BGP.

Stratum (running on each switch) and ONOS (running off-switch and managing a network of switches) communicate using the following interfaces.

- **P4:** Defines the forwarding behavior for programmable switching chips as well as model fixed-function ASIC pipelines. A P4 program defines a contract that is implemented by the data plane and programmable by the control plane.

- **P4Runtime:** An SDN-ready interface for controlling forwarding behavior at runtime. It is the key for populating forwarding tables and manipulating forwarding state, and it does so in a P4 program and hardware agnostic way.

- **OpenConfig Models:** Define a base for device configuration and management. These models can be programmatically extended for platform-specific functionality, but the goal is to minimize model deviations so as to enable a vendor-agnostic management plane.

- **gNMI** (gRPC Network Management Interface): Improves on the existing configuration interfaces by using a binary representation on the wire and enabling bi-directional streaming. Paired with the OpenConfig models, gNMI is SDN-ready.

- **gNOI** (gRPC Network Operations Interfaces): A collection of microservices that enable switch specific operations, like certificate management, device testing, software upgrade, and networking troubleshooting. gNOI provides a semantically rich API that replaces existing CLI-based approaches.

Trellis, as an SDN control application running on top of ONOS, controls packet forwarding across the switching fabric internal to a CORD POD (i.e., within a single site). But Trellis can also be extended across multiple sites deeper into the network using multiple stages of spines, as shown in Figure 6.3. This means Trellis has the potential to play a role in implementing the backhaul and midhaul network for the RAN, or alternatively, extending the RAN into customer premises (denoted "On Site" in the figure).

The software stack we've just described is substantial, and has the potential to disrupt and transform the Internet in ways that can only be matched by 5G. Of particular note, the RAN Intelligent Controller shown in Figure 6.1 is implemented as a set of extensions to ONOS. This puts the ONOS-based RIC at the very center of the design, where the SDN and 5G worlds intersect.

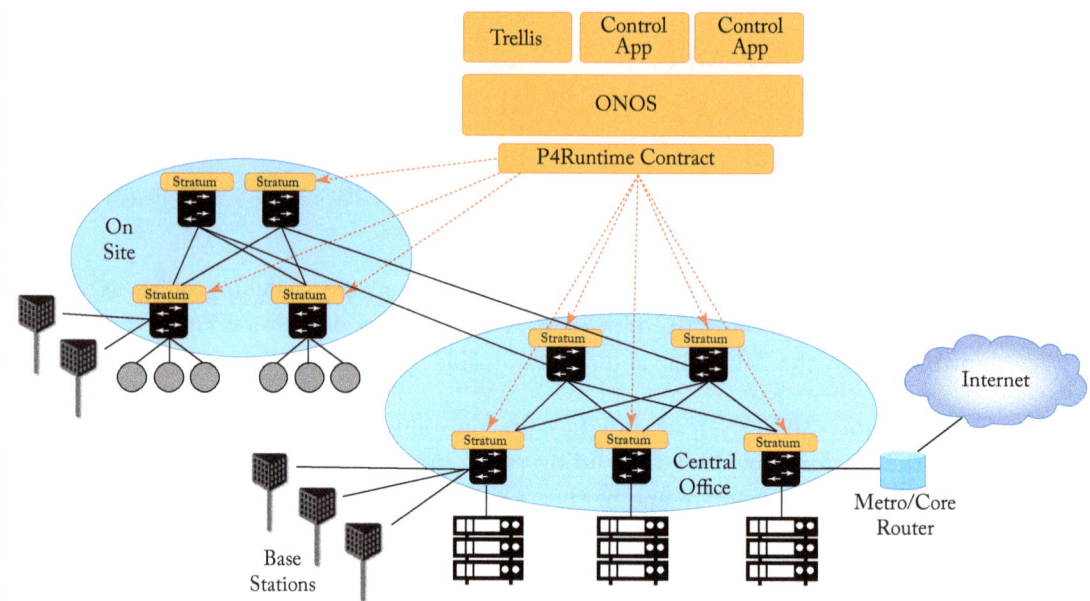

Figure 6.3: Trellis control application managing a (possibly distributed) leaf-spine fabric.

Further Reading

For more information about the SDN software stack, we recommend a companion book: Software-Defined Networks: A Systems Approach. March 2020.

CHAPTER 7

Cloudification of Access

The previous chapters went step-by-step, first breaking 5G down into its elemental components and then showing how those components could be put back together using best practices in cloud design to build a fully functional, 3GPP-compliant 5G cellular network. In doing so, it is easy to lose sight of the big picture, which is that the cellular network is undergoing a dramatic transformation. That's the whole point of 5G. We conclude by making some observations about this big picture.

7.1   MULTI-CLOUD

To understand the impact of cloud technologies and practices being applied to the access network, it is helpful to first understand what's important about the cloud. The cloud has fundamentally changed the way we compute, and more importantly, the pace of innovation. It has done this through a combination of the following.

- **Disaggregation:** Breaking vertically integrated systems into independent components with open interfaces.

- **Virtualization:** Being able to run multiple independent copies of those components on a common hardware platform.

- **Commoditization:** Being able to elastically scale those virtual components across commodity hardware bricks as workload dictates.

There is an opportunity for the same to happen with the access network, or from another perspective, for the cloud to essentially expand so far as to subsume the access network.

Figure 7.1 gives a high-level overview of how the transformation might play out, with the global cloud spanning edge clouds, private Telco clouds, and the familiar public clouds. Each individual cloud site is potentially owned by a different organization (this includes the cell towers, as well), and as a consequence, each site will likely be multi-tenant in that it is able to host (and isolate) applications on behalf of many other people and organizations. Those applications, in turn, will include a combination of the RAN and Core services (as described throughout this book), Over-the-Top (OTT) applications commonly found today in public clouds (but now also distributed across edge clouds), and new Telco-managed applications (also distributed across centralized and edge locations).

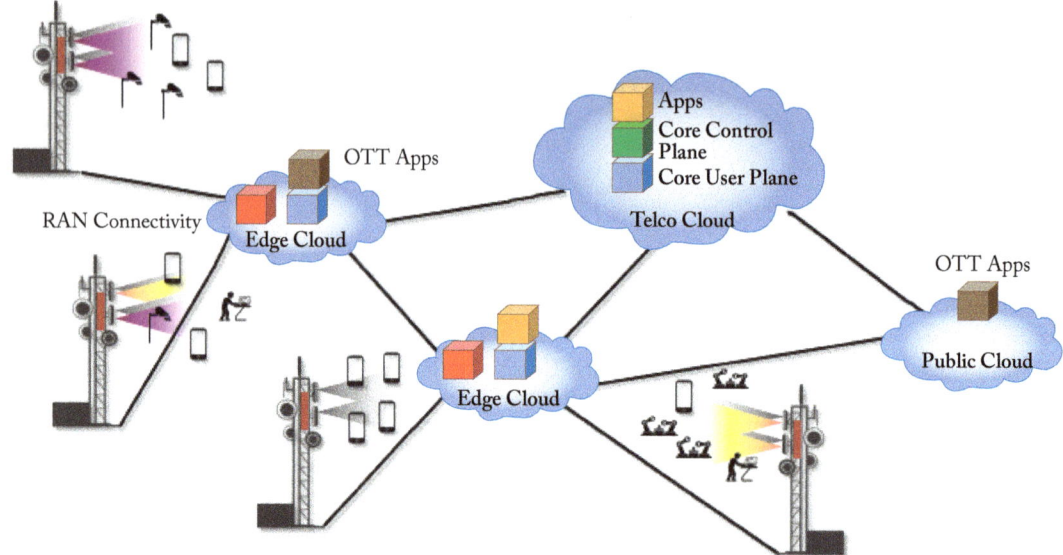

Figure 7.1: A multi-tenant/multi-cloud—including virtualized RAN resources alongside conventional compute, storage, and network resources—hosting both Telco and Over-the-Top (OTT) services and applications.

Eventually, we can expect common APIs to emerge, lowering the barrier for anyone (not just today's network operators or cloud providers) to deploy applications across multiple sites by acquiring the storage, compute, networking, and connectivity resources they need.

7.2 EDGECLOUD-AS-A-SERVICE

Of all the potential outcomes discussed in the previous section, one that is rapidly gaining traction is to run a 5G-enabled edge cloud as a centrally managed service. As illustrated in Figure 7.2, the idea is to deploy an edge cloud in enterprises, configured with the user plane components of the RAN and Mobile Core (along with any edge services the enterprise wants to run locally), and then manage that edge deployment from the central cloud. The central cloud would run a management portal for the edge cloud, along with the control plane of the Mobile Core. This is similar to the multi-cloud configuration discussed in Section 5.2, except with the added feature of being able to manage multiple edge deployments from one central location.

The value of such a deployment is to bring 5G wireless advantages into the enterprise, including support for predictable, low-latency communication required for real-time controlling of large numbers of mobile devices. Factory automation is one compelling use case for such an edge cloud, but interest in supporting IoT in general is giving ECaaS significant momentum.

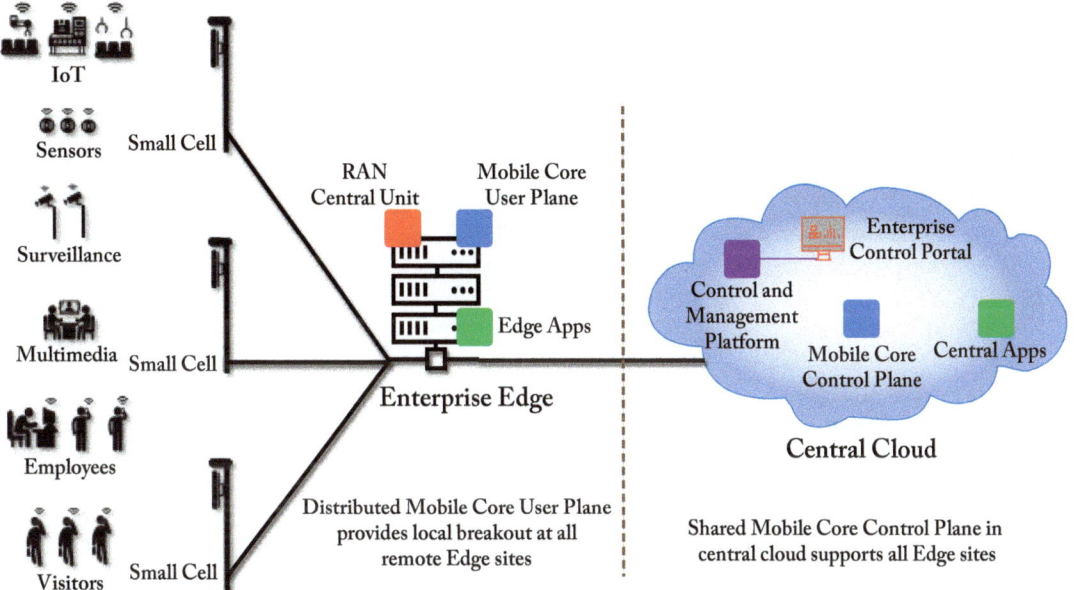

Figure 7.2: EdgeCloud-as-a-Service, a managed service, with RAN and Mobile Core user plane components running in the enterprise, and the control plane of the Mobile Core (along with a management portal) running centrally in the public cloud.

This momentum has, not surprisingly, led to recent commercial activity. But there is also an open source variant, called Aether, now available for early adopters to evaluate and experiment with. Aether is an ONF-operated ECaaS with 4G/5G support, built from the open source components described throughout this book. Aether works with both licensed and unlicensed frequency bands (i.e., CBRS), but it is the latter that makes it an easy system to opt into. Figure 7.3 depicts the early stages of Aether's centrally managed, multi-site deployment.

Note that each edge site in Figure 7.3 corresponds to a CORD POD described in Chapter 6, re-configure to off-load the O&M Interface and the Control elements of the Mobile Core to the central cloud.

Further Reading

For more information about Aether, visit the Aether Web Site. ONF, March 2020.

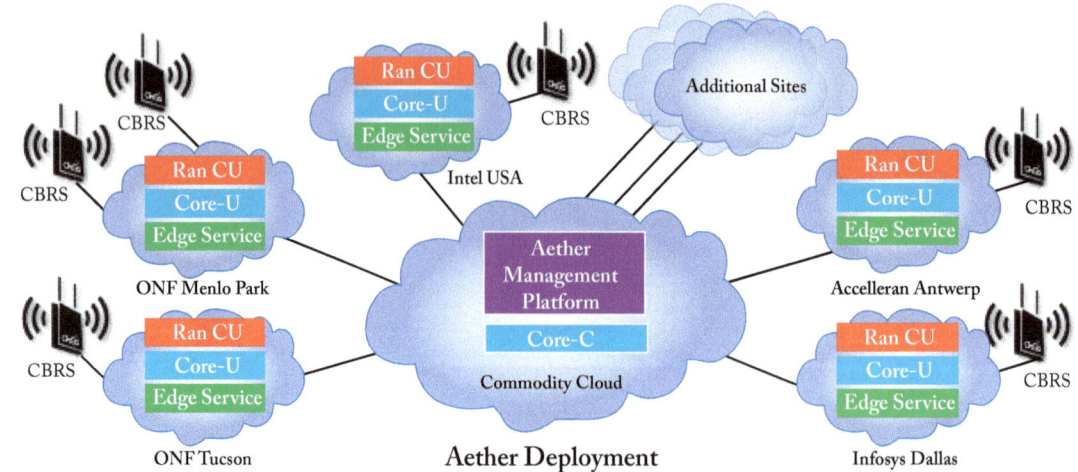

Figure 7.3: Aether is an ONF-operated EdgeCloud-as-a-Service built from the SD-RAN and disaggregted Mobile Core components described throughout this book. Aether includes a centralized operations portal running in the Public Cloud.

7.3   RESEARCH OPPORTUNITIES

In order for the scenarios described in this chapter to become a reality, a wealth of research problems need to be addressed, many of which are a consequence of the blurring line between access networks and the edge cloud. We refer to this as the *access-edge*, and we conclude by identifying some example challenges/opportunities.

- **Multi-Access:** The access-edge will need to support multiple access technologies (e.g., WiFi, 5G, fiber), and allow users to seamlessly move between them. Research is needed to break down existing technology silos, and design converged solutions to common problems (e.g., security, mobility, QoS).

- **Heterogeneity:** Since the access-edge will be about low-latency and high-bandwidth connectivity, much edge functionality will be implemented by programming the forwarding pipeline in white-box switches, and more generally, will use other domain-specific processors (e.g., GPUs, TPUs). Research is needed to tailor edge services to take advantage of heterogeneous resources, as well as how to construct end-to-end applications from such a collection of building blocks.

- **Virtualization:** The access-edge will virtualize the underlying hardware using a range of techniques, from VMs to containers to lambdas, interconnected by a range of L2, L3, and L4/7 virtual networks, some of which will be managed by SDN control applications. Research is needed to reconcile the assumptions made about by cloud native services and access-oriented Virtualized Network Functions (VNFs) about how to virtualize compute, storage, and networking resources.

- **Multi-Tenancy:** The access-edge will be multi-tenant, with potentially different stakeholders (operators, service providers, application developers, enterprises) responsible for managing different components. It will not be feasible to run the entire access-edge in a single trust domain, as different components will operate with different levels of autonomy. Research is needed to minimize the overhead isolation imposed on tenants.

- **Customization:** Monetizing the access-edge will require the ability to offer differentiated and customized services to different classes of subscribers/applications. Sometimes called network slicing, this involves support for performance isolation at the granularity of service chains—the sequence of functional elements running on behalf of some subscriber. Research is needed to enforce performance isolation in support of service guarantees.

- **Near-Real Time:** The access-edge will be a highly dynamic environment, with functionality constantly adapting in response to mobility, workload, and application requirements. Supporting such an environment requires tight control loops, with control software running at the edge. Research is needed to analyze control loops, define analytic-based controllers, and design dynamically adaptable mechanisms.

- **Data Reduction:** The access-edge will connect an increasing number of devices (not just humans and their handsets), all of which are capable of generating data. Supporting data reduction will be critical, which implies the need for substantial compute capacity (likely including domain-specific processors) to be available in the access-edge. Research is needed to refactor applications into their edge-reduction/backend-analysis subcomponents.

- **Distributed Services:** Services will become inherently distributed, with some aspects running at the access-edge, some aspects running in the datacenter, and some running on premises or end device (e.g., on-vehicle). Supporting such an environment requires a multi-cloud solution that is decoupled from any single infrastructure-based platform, with research needed to develop heuristics for function placement.

- **Scalability:** The access-edge will potentially span thousands or even tens of thousands of edge sites. Scaling up the ability to remotely orchestrate that many edge sites (even at just the infrastructure level) will be a qualitatively different challenge than managing a single datacenter. Research is needed to scale both the edge platform and widely deployed edge services.

Further Reading

To better understand the research opportunity at the access-edge, see Democratizing the Network Edge. ACM SIGCOMM CCR, April 2019.

Authors' Biographies

LARRY PETERSON

Larry Peterson is the Robert E. Kahn Professor of Computer Science, Emeritus at Princeton University, where he served as Chair from 2003–2009. He is a co-author of the best selling networking textbook *Computer Networks: A Systems Approach* (6e), which is now available as open source on GitHub. His research focuses on the design, implementation, and operation of Internet-scale distributed systems, including the widely used PlanetLab and MeasurementLab platforms. He is currently working on a new access edge cloud called CORD, an open source project of the Open Networking Foundation (ONF), where he serves the CTO.

Professor Peterson is a former Editor-in-Chief of the *ACM Transactions on Computer Systems* and served as program chair for SOSP, NSDI, and HotNets. He is a member of the National Academy of Engineering, a Fellow of the ACM and the IEEE, the 2010 recipient of the IEEE Kobayashi Computer and Communication Award, and the 2013 recipient of the ACM SIGCOMM Award. He received his Ph.D. from Purdue University in 1985.

OĞUZ SUNAY

Oğuz Sunay is currently the Vice President for Research & Development at ONF, where he leads all mobile-related projects. Prior to that, he served as the Chief Architect for Mobile Networking at ONF. Before joining ONF, Oğuz was the CTO at Argela-USA, where he was the innovator of a Programmable Radio Access Network Architecture (ProgRAN) for 5G that enabled the world's first dynamically programmable RAN slicing solution. He has also held prior industry positions at Nokia Research Center and Bell Laboratories, where he focused on 3G and 4G end-to-end systems architectures and participated and chaired various standardization activities. Oğuz has also spent over 10 years in academia, as a professor of electrical and computer engineering. He holds many U.S. and European patents on various aspects of 3G, 4G, and 5G, and has authored numerous journal and conference publications. He received his Ph.D. and M.Sc. from Queen's University, Canada, and his B.Sc.Hon. from METU, Turkey.

Lightning Source UK Ltd.
Milton Keynes UK
UKHW051956160421
382090UK00003B/4